A RE-APPRAISAL OF FORESTRY DEVELOPMENT IN DEVELOPING COUNTRIES

FORESTRY SCIENCES

Also in this series:

Prins CFL ed: Production, Marketing and Use of Finger-Jointed Sawnwood.
  ISBN 90-247-2569-0
Oldeman RAA et al. eds: Tropical Hardwood Utilization: Practice and
  Prospects. 1982, ISBN 90-247-2581-X
Baas P ed: New Perspectives in Wood Anatomy. 1982. ISBN 90-247-2526-7
Bonga JM and Durzan DJ eds: Tissue Culture in Forestry. 1982.
  ISBN 90-247-2660-3
Satoo T and Madgwick HAI: Forest Biomass. 1982. ISBN 90-247-2710-3
Den Ouden P and Boom BK eds: Manual of Cultivated Conifers: Hardy in
  Cold- and Warm-Temperature Zone. 1982. Hardbound ISBN 90-247-2148-2;
  paperback ISBN 90-247-2644-1
Van Nao T ed: Forest Fire Prevention and Control. ISBN 90-247-3050-3

In preparation:

Gordon JC and Wheeler CT eds: Biological Nitrogen Fixation in Forest
  Ecosystems: Foundation and Applications
Hummel FC ed: Forestry Policy
Németh MV: The Virus – Mycoplasma and Rickettsia Disease of Fruit Trees
Powers RF and Miller HG eds: Applied Aspects of Forest Tree Nutrition
Powers R.F. and Miller HG eds: Basic Aspects of Forest Tree Nutrition
Rajagopal R: Information Analysis for Resource Management

A Re-Appraisal of Forestry Development in Developing Countries

by

JAMES J. DOUGLAS
Bureau of Agricultural Economics
Canberra, Australia

1983 **MARTINUS NIJHOFF / DR. W. JUNK PUBLISHERS**
a member of the KLUWER ACADEMIC PUBLISHERS GROUP
THE HAGUE / BOSTON / LANCASTER

Distributors

for the United States and Canada: Kluwer Boston, Inc., 190 Old Derby Street, Hingham, MA 02043, USA
for all other countries: Kluwer Academic Publishers Group, Distribution Center, P.O.Box 322, 3300 AH Dordrecht, The Netherlands

Library of Congress Cataloging in Publication Data

```
Douglas, James J.
  A re-appraisal of forestry development in developing
countries.

  (Forestry sciences ; v. 8)
  Bibliography: p.
  1. Underdeveloped areas--Forestry--Economic aspects.
2. Forests and forestry--Economic aspects--Bangladesh.
I. Title.  II. Series.
HD9750.5.D68  1983     333.75'15'091724      83-3989
ISBN 90-247-2830-4
```

ISBN 90-247-2830-4 (this volume)

Copyright

© 1983 by Martinus Nijhoff/Dr W. Junk Publishers, The Hague.

All rights reserved. No part of this publication may be reproduced, stored in a retrieval system, or transmitted in any form or by any means, mechanical, photocopying, recording, or otherwise, without the prior written permission of the publishers,
Martinus Nijhoff/Dr W. Junk Publishers, P.O. Box 566, 2501 CN The Hague, The Netherlands.

PRINTED IN THE NETHERLANDS

CONTENTS

Preface vii

PART 1: ISSUES IN ECONOMIC DEVELOPMENT 1

Chapter 1: Economic Growth, Population and Modernisation 2
    1.1 Growth and the GNP Criterion 3
    1.2 Population and Development 5
        1.2.1 Some general observations on population 6
        1.2.2 The case of Bangladesh 11
    1.3 Modernisation Theories and the Allocation of Capital 14
        1.3.1 The take-off to development and the 'Big Push' 14
        1.3.2 Agriculture re-instated 18
        1.3.3 A case in point 19

Chapter 2: Income Distribution: The Critical Issue 33
    2.1 Trade, Tariffs and the New International Economic Order 33
    2.2 Financial Dualism, the Monetary System and Disguised Unemployment 44
    2.3 Development and Income Distribution 45
        2.3.1 Dependency theory 55
        2.3.2 Appropriate technology 58
        2.3.3 Basic needs 62

PART 2: THE ROLE OF FORESTRY IN DEVELOPMENT 66

Chapter 3: Approaches to Forestry Development 67
    3.1 Project Evaluation in Forestry 68
    3.2 Forestry in Development: The Original Westoby View 71
    3.3 The Forest Industries of Bangladesh 74
        3.3.1 Pulp and Paper 75
        3.3.2 Manufactured Board 76
        3.3.3 Sawmilling 78
        3.3.4 Interpretations 78
    3.4 Is There a Forest Industry Option? 79
    3.5 Forestry in Development: Newer Approaches 83
    3.6 The Malaysian Example 93

Chapter 4: Forests, Fuelwood and People: The Linkage 96
    4.1 Tropical Forests and the Environment 97
        4.1.1 The size and influence of the forest 97
        4.1.2 The use of the forest 99
    4.2 The Fuelwood Question 104
        4.2.1 The general situation 104
        4.2.2 Fuelwood and the rural poor in Bangladesh 108

4.3	Forests and Rural Development	115
	4.3.1 The Philippines: a small farmer tree growing project	116
	4.3.2 Northern Thailand	118
	4.3.3 Kenya	119
	4.3.4 China	119
	4.3.5 Bangladesh: a community forestry project	120
4.4	Is Agro-forestry Always the Answer?	124

PART 3:	CONSTRAINTS AND CONCLUSIONS	126
Chapter 5:	New Approaches to Development	127
	5.1 Some Political and Historical Aspects	127
	5.2 A Perspective on the Economics of Development	130
Chapter 6:	Forestry and Development: the Problem Re-stated	134

REFERENCES		143
APPENDIX A:	The Economy of Bangladesh	152
APPENDIX B:	Forest Industries of Bangladesh	165
APPENDIX C:	Conversion of Plain Land Forests of Bangladesh to Fast-Growing Plantation	174

PREFACE

This book is directed at foresters who work, or have an interest, in the developing world, and at development analysts and theorists who are concerned with the forestry sector. Most readers will be aware that in recent years, some fundamental changes in thinking about the development process in very poor countries have occurred. At one level, the underdevelopment problem has been explained as a lack of absorptive capacity, or implementation ability in very poor countries. However, it now seems that these are only symptoms of a more profound ailment in the whole economic structural and philosophical approach to development. The idea that poor countries could transform their economies through an accelerated process of industrialisation has proved largely incorrect, or at least highly premature. Within the rural sector, emphasis on productivity and aggregate income growth have been shown to have had little effect or, worse still, negative effects, on the burgeoning group of poor and landless rural dwellers.

The implications of these observations are no less important for forestry than for other sectors. Foresters may be tempted to take comfort in the idea that their sector is <u>rural</u>, and therefore exempt from the distributional and structural arguments which have been levelled against other sectors of LDC economies. But, where forestry is practiced as a large scale land using exercise, with provision of raw materials for capital intensive, high technology processing industries as its main purpose, this classification in the LDC context becomes irrelevant. In fact, the justification for forestry of this type may actually be even more suspect than the equivalent approach in other sectors, because it occupies land, which is still the most basic productive resource in these countries.

One of my principal objectives in this book will be to examine the extent to which forest economists, analysts and administrators have responded to the new directions in development thinking. I will be arguing that so far as the <u>practice</u> of forestry in LDC's is concerned, the response so far has been limited, and indeterminate.

I believe there is a place for a book which tries to form the linkage between general economic issues in development, and the specific concerns and characteristics of the forestry sector. Certainly, at the practical level in the country from which I have drawn many examples, it is precisely in this area between the conception and setting of macro-economic objectives on the one hand, and the design of sector programmes and projects on the other, where critical breakdowns in planning occur. I also believe I am on safe ground in asserting that this is a fairly common problem in LDC's. It is perhaps too easy for development economists, no matter how clearly they may see the overall directions an economy should take, to regard the translation of this into specific sectoral programmes as someone else's problem. And it is too tempting for foresters, faced with such olympian detachment on the part of economists, to become engrossed in the technicalities of on-going sectoral programmes, rather than begin the arduous task of overhauling the conceptual base, objectives and <u>modus operandi</u> of the sector. Indeed, in the absence of more effective communication between maroeconomic planners, and forestry sector administrators it is perfectly possible that the latter may not even understand, or agree, that their programmes, based as they predominantly are on outmoded and inapplicable concepts of development, <u>are</u> out of phase with the general economic strategy for the country concerned.

As will already be apparent, this exposition is polemical in style. It seeks to present a definite viewpoint, because my objective is to alter the way people think about forestry and economic development, rather than to provide an academic review of the subject (albeit that there is a need for the latter). Thus, whilst I have tried to cover the theoretical and empirical ground, I have done so in some cases only by briefly mentioning what may be large, complex issues. There is no alternative to this if the book is to be kept in reasonable bounds. In general, I have referred to published studies which I believe to be seminal, or typical, for the issues and material I believe to be relevant, in preference to attempting an exhaustive inventory of it.

I have used examples throughout the book, although I hope that the general argument presented has sufficient force to stand alone, rather than on the basis of case studies. The reason for their inclusion is really as a demonstration of the variability of the situations that prevail in different countries. Many of my specific references will deal with various aspects of the situation in Bangladesh, in a country I have had the opportunity to study in some detail. I believe there is some point in this rather more determined attempt to follow through the application of general theoretical observations to a particular case, than is often done in the development literature. It will illustrate, I suggest, the <u>level</u> of information and observation of a particular country situation needed, before useful and practical conclusions can be drawn.

The book is arranged into three basic parts. In the first, general concepts and issues in economic development are considered. In the second, development concepts are used to analyse the role and nature of the forestry sector. A third, smaller part attempts to draw together the observations made in the two major sections into a coherent re-statement of findings and conclusions.

To Silvia

PART 1

ISSUES IN ECONOMIC DEVELOPMENT

"... money. They didn't want to talk about it. They didn't want to face how much they cared and worried about it. They wished to pretend - to themselves as much as to others - that it affected them very little. ... Like most hypocrisies it had something good in it. They wanted to disguise their motives, because they would have liked loftier ones.... Most men thought about money more than they admitted, and badly wanted it. The only thing they wanted more, perhaps, was that other men shouldn't have more money than themselves: or ideally have less".

C.P. Snow (<u>In Their Wisdom</u>)

Chapter 1

ECONOMIC GROWTH, POPULATION AND MODERNIZATION

Economic development; what it is, what causes it, and what inhibits it, is an inexhaustible vehicle for theorizing and, usually at some years remove, for practical experimentation in countries deemed to be in need of it. It is a reasonable, if not particularly encouraging generalization to observe that many of the physical, material manifestations of post War development theories around the world have persisted a good deal longer than the theories themselves. The early chapters of this book will consider the basic theories involved, and will seek to broadly answer some basic questions: what is development, and under-development? What causes major economic shifts in poor countries? How will these effect the various socio-economic groupings in those countries?

To reiterate a point made in the Introduction: all the major issues raised below are treated rather briefly, and some are treated more briefly than others. There is no alternative to this approach, if this exposition is to be kept to digestible proportions. However, brevity should not be taken as being indicative of unimportance: the issues raised are all important, and they are all interdependent to some extent, albeit that they are discussed in separate categories below.

The reasons for attempting an overview of this very large subject of development issues will be clear enough: whenever an LDC government takes a decision to implement a given project or programme - in forestry of anywhere else - and/or when a multi-lateral or bi-lateral development assistance agency provides certain types of material or technical inputs, acceptance of a certain <u>concept</u> of development is already implicit. The development model into which the scheme fits may be fully specified, or merely guessed at, or not even considered by the operatives involved, but it will be there nonetheless. The success or failure of the scheme will depend, in far larger measure than may once have been thought, on whether the model of development into which it fits is appropriate to, <u>and</u> operative in the particular country in question.

1.1 Growth and the GNP criterion

The simplest definition of development is the attainment of a condition where the nation concerned is recording significant and sustained economic growth. In the 1960s and early 1970s, the growth objective has come in for a deal of criticism. In the developed world, criticism has centred on the unpleasant consequences of quite massive economic growth. In these rich countries, however, there seems to have been little success with alternatives to growth based models, and the area of debate has shifted to the qualitative aspects of growth: ecological concerns: the energy balance; and so on. More recently still, a certain nostalgia for the high growth era - whatever its ecological consequences - has set in, and can be expected to remain at least until more substantial rates of growth return.

In the developing world, the essential debate has been on ways and means to link up the growth criterion with broader social concepts of development.

The material progress of a nation is usually expressed in terms of Gross National Product, or variants of it. The GNP criterion suffers from a number of limitations:

(1) In its unadjusted form it pays relatively little attention to the cost of pollution and other externalities related to quality of life - this is the essence of the debate (referred to above) which has been going on in developed countries.

(2) More importantly from the viewpoint of measuring development in poorer countries, it is a monetary index, and is often expressed for comparative purposes in terms of a major international hard currency. There are in fact two related problems with this. Firstly, the degree of penetration of the money system of exchange into the traditional barter economy varies widely amongst LDC's. As a measure of the purchasing power of individuals or of their level of consumption, it can consequently be highly inaccurate. Even for commodities which are heavily traded internationally - such as agricultural products - their domestic prices within LDC's are

frequently quite different to those prevailing in international markets. The second problem in this category is the unreality of expressing economic welfare in a given LDC in terms of its currency as valued against some international standard. There are many reasons for currency fluctuations, and clearly, the fact that a given LDC's currency may move substantially against the internatonal parity generally does not imply a correspondingly large shift in the living standard or domestic consumption level of that country. Yet, across the board international economic comparisons based on GNP usually makes no allowance for this reality. It is quite possible, given these circumstances, for a given LDC's currency to be overvalued in terms of its international purchasing power yet at the same time for its GNP to understate the living standard of its population. Croswell (1981) refers to work by Kravis et al which has attempted to deal with such disparities by using an exchange rate deviation index which measures the extent to which nominal comparisons with income in the United States misrepresents real comparisons.

(3) GNP is an aggregate income criterion, and of itself says nothing of how wealth is distributed in the country concerned. Some years ago a major conflict in development circles arose around the question of whether aggregate income growth is the only important criterion in development: whether, once growth was in progress, all desired social and distributional changes would automatically follow. Much of what follows in this book relates to this fundamental question, and it will be discussed specifically in Chapter 2: there is now a strong body of opinion which suggests that sustained development in a poor country will not occur in any meaningful sense, unless specifically targetted relief to the characteristically large group of destitute or very poor people is forthcoming. Moreover, some recent empirical results indicate that in fact there may be much more of a _confluence_ between maximizing aggregate returns to capital investment in LDC's, and redistributing income (or income opportunities) towards the poorest groups in these societies; the idea of there being an inexorable trade-off between the two objectives may not apply.

1.2 Population and development

The matter of population growth has occupied the minds of development theorists and field operatives at least as much as any other single issue, and analyses of the subject have a long history. For all that, it remains contentious and imperfectly understood. The wide range of possible interpretations of historical demographic trends, and the perennial difficulty of distinguishing cause from effect in this area, has blurred even the clarity of hindsight.

The basic doctrine of Malthus - the principle of the stationary state - is well known. Malthus suggested that in an agricultural society, the trend of population would lead to ever increasing pressure on the finite land base: rents would rise, wages would fall and society would be locked into an equilibrium of basic survival and misery. To Malthus' simple and depressing theorem, refinements and variations have been added: since rapid population growth forces more people into lower income strata, savings drop, and the potential of the country to attain minimum savings levels for the take-off to development out of the agriculture sector (of which more later) will be reduced. The physical capital of workers themselves will be reduced. The need for more public infrastructure will siphon off funds from more 'productive' ends. Pressure on foreign resources, through the need to import food and other basic necessities, may force nations to invest in uneconomic import-saving industries. High rates of population growth will lead to larger proportions of the population below useful working age, adding to the consumption burden of the workforce. The list of unpleasant ramifications seems, from a reading of literature of this type, endless.

The interpretation of what occurred in developed nations during their developmental phase by protagonists of this view is more or less that the nations of Western Europe, for example, began with lower rates of population growth than present day LDC's, and the fall in death rates over a long period was balanced by declining fertility - both of which were a result of the development process itself. Thus, by this interpretation, the Western countries have come to their extremely low rates of population growth. By contrast, it is suggested, developing countries have very high rates of population growth. Mortality rates have dropped through transfers of health technology, but there is an absence of the various socio-economic changes that produce a corresponding fall in birth rates.

Meier (1976), in discussing this view of the situation, suggests that the relevant questions become: can birth control measures change this situation, without waiting for the socio-economic factors to change? Is the future of population essentially a question of development? Or is the reverse true? Meier concludes that rapid population is a cause, as well as a consequence of poverty, and the beneficial effects on development of declining fertility should be recognised.

1.2.1 Some general observations on population

Before moving directly to a discussion of the major questions, it will be useful to consider the recent (or, more accurately perhaps, the recently discovered) phenomenon of widespread fertility declines in LDC's. In 1968, the United Nations published the following estimates of past, present and future levels of population:

Table 1.1: WORLD POPULATION (IN MILLIONS) 1920-2000

	1920	1940	1960	1980	2000
More development countries	606	730	854	1 052	1 266
Less development countries	1 256	1 565	2 136	3 415	5 248
Total	1 862	2 295	2 990	4 467	6 504

Source: World Population Prospects, TAI, UN. 1968.

The relatively pessimistic outlook on population implied in these figures seemed to be borne out by figures recorded throughout the 1960s. The U.N. (1974) observed that the growth rate of the world population during the 1960s had been 2 per cent per annum; higher than ever before. Various attempts were made to analyse the causes, and provide solutions. The U.N. itself, in the same publication, entered the field with: 'the processes of economic and social development and the demographic transition (to lower birth rates) are linked in a mutually causative fashion, death rates being reduced by improved living standards made possible by an economic diversification process which, in its turn, causes - and is, in varying degree, contingent upon - a reduction in fertility'. To the extent that this can be interpreted as meaning anything at all, it seems to be a variant of the classic, if somewhat unsatisfying economic maxim that everything depends on everything else.

At this stage, the gross figures available seemed to bear out the arguments that changes in per capita income growth were in inverse relationship with population growth. The U.N. argued further that the same inverse relationship existed between savings, education and nutrition on the one hand, and rates of change of population on the other. However, even at this stage, these relationships may not have been as firm as implied. The U.N. figures themselves in fact show that for the variables of per capita GDP growth, and per capita calorific and protein intake, countries in the +3 per cent population growth category had performed as well, or better, than those in the 2.00-2.49 per cent and 2.50-2.99 per cent categories.

By the early 1970s, it began to become apparent that a fertility transition, from high rates to significantly lower rates of birth, was under way in many LDC's. For quite a few of them it had been in progress for some years, but had been concealed by lags or aggregations in regional population figures; a great many countries seemed to have entered this transitional state fairly soon after World War II. Kirk (1971), among the first to identify the scope of the transition, noted that it seemed to be progressing rapidly, and that the period of transition had shortened very considerably when compared to the historical example of Western nations. Perhaps the simplest illustration of this is the total births figure for 1977, which according to U.N. and USAID data cited by Eberstadt (1980), was 121.3 million. However, application of the older U.N. projections to this year (using interpolation) produced figures around 144-145 million, and this demonstrates the scale of the fertility shift that has occurred.

Eberstadt collates some comparative data on vital rates since the 1950s which identify more closely the nature of recent changes, and Table 1.2 below extracts from these data.

Table 1.2: CHANGES IN WORLD VITAL RATES, 1950-1977

Births (per 1000)	1950	1960	1970	1977
Developed countries	23	21	18	16
China	40	37	30	26
Other LDC's	44	43	42	37
Deaths (per 1000)				
Developed countries	10	9	9	9
China	23	16	11	10
Other LDC's	26	19	17	15

Source: Eberstadt (1980) Table 2.

These figures show the relatively significant drop in death rates between 1950 and 1960. Even more dramatic are the falls in birth rates after 1970, but the 'bunching' of this effect to such recent years explains why the transition was missed previously.

Whether or not this phenomenon will be sustained, and what levels of fertility it will lead to, depend to a large extent on what assessment of the causes of the change is accepted, and what is predicted for those factors. Here, the subject becomes somewhat complicated. Eberstadt (op. cit.) provides some very interesting figures derived from Sivard which show for a group of 12 large LDC's that none of the socio-economic indicators which might be expected to be related to the birth rate (with the exception of literacy) are correlated with it at a statistically significant level. That is, none of: per capita GNP; infant mortality; life expectancy; per capita calorific intake; or percentage of university students who are female; are significant factors (according to Eberstadt) despite the fact that other studies have accepted them as being so. Eberstadt suggests 'macro' theories of change are missing the complexity of the process of fertility decline, or are aimed at the wrong place; the national, rather than the family level.

A possible explanation of this apparent contradiction to other
results (apart from the obvious difference in time frame) might be in
method. Sivard's figures are essentially a cross-sectional comparison
across widely differing nations. Even if the various factors analysed
were operating on birth rates in these countries, they would not be
operating from the same base levels or in the same proportions. It is
notoriously difficult to obtain significant results in cross-sectional
studies using economic and social parameters, for this sort of reason.
Eberstadt himself comments on the general diversity of countries in the
LDC group and, given this, it is hardly surprising that inter-country
comparisons yield little. It will be most interesting to obtain the
results of within country correlation tests, using time series data for
the rapid decline period, of birth rate against the 'indicator'
variables, but it may be some years before sufficient observations from
the fertility decline period are available. Mauldin and Berelson (1978)
have utilized 1965-1975 data to provide some fairly broad threshold
ranges for major socio-economic variables; the ranges given are those
regarded as conducive to fertility declines.

Table 1.3: INDICATORS OF CONDITIONS FAVOURABLE TO FERTILITY DECLINES

Variable	Threshold Range
	%
Population in cities 20 000 +	16-50
Non-agricultural labour force	50-65
Life expectancy	60-70
Female marriage before 20	10-29
Female literacy	60-75
Hospital beds/1000 population	5
Newspaper circulation/1000 population	70-100

Source: Mauldin and Berelson (1978).

It would certainly not be prudent at this stage to write off the
explanatory powers of socio-economic variables at the national level.
Nor, however, should Eberstadt's plea for a search for causes at the
family level be discarded. He suggests that changing educational
opportunities and aspirations in many developing countries, and other
reversals of inter-generational wealth flows (so that children tend to

become net gainers in welfare, rather than net contributors to the family as they are in purely agrarian societies) need to be closely examined as causes. Changes in the status of women (who, Eberstadt claims, generally desire fewer children then men) is also a potentially important factor.

Much of the debate in the literature on population centres around the efficacy and ethics of various methods of birth control. This is an interesting question, but tends to be highly country-specific: accordingly, we will consider it in the context of a specific country towards the end of this section. The interest here is in an overall interpretation of what has happened, or is happening in population growth in LDC's, before we attempt an interpretation of a given case. Eberstadt makes the general observation that birth control programmes rarely succeed in altering the will of people in the population to have certain numbers of children; they can assist in reducing the number of unwanted children. And it is certainly true that virtually all people limit reproduction (one way or another) to below the maximum limit they could produce. It is reasonable to argue, therefore, that changes in socio-economic conditions and social perceptions, whether at the family or national level, will be at least as important as introduction of birth control education and techniques. However, at this stage it is rather more difficult to go beyond this observation than might once have been thought. It is apparent that high rates of population growth can co-exist with high levels of per capita income growth; the Malthusian notion of an eventual slowdown in population growth through 'immiserisation' (i.e., higher death and disease rates) - an idea also popular with some Marxist theoreticians - has not often been borne out in fact. Slowdowns have occurred in an environment of <u>falling</u> death rates and increased life expectancies in LDC's.

For all the above reasons, at this stage some caution should be exercised when attempting to predict what sorts of changes might occur in overall population levels in given countries, on the basis of other measurable economic factors, and in attempting to estimate the potential efficacy of population limitation programmes. Finally, general conclusions as to the nexus between population growth and human welfare, or development and declining fertility, should be avoided. On these points, what is known - even what has occurred in the past - is open to varied interpretation: earlier in this section, the common interpretation of a relationship between development and fertility decline in 19th Century Western Europe was noted. However, the fact is that fertility declined in England and France for sustained periods in the 19th Century when living standards were not rising; when, in fact, they were declining. In Japan, by contrast, living standards began to improve appreciably in the late 19th Century, but there was no decline in fertility until the catastrophic events of the Second World War.

1.2.2 The case of Bangladesh

As noted earlier in this section, when discussing practices and results of population control, we are well advised to deal in specific examples. Bangladesh provides a useful case study, if only because of the extreme nature of the population problems in that country: the density of population in Bangladesh now exceeds 600 persons per square kilometre, and the rate of growth of that population has - at least until very recently - remained stubbornly high. In a very real sense, for Bangladesh and countries like it, population is the parameter against which all other social and economic indicators must be measured. Faaland and Parkinson (1976) say:

> 'The growth of population is a threat which if allowed to continue unchecked, could eventually bring the economy back - even after initial improvements in living standards - to a Malthusian State'

They estimate the population of Bangladesh in 1975 as 80 million, and they provide the following projections:

Table 1.4: POPULATION PROJECTIONS FOR BANGLADESH
(in millions)

1975	1980	1985	1990	1995	2000
80	93	106	121	136	150

Source: Table VI, Faaland and Parkinson (1976).

These figures imply an annual rate of increase of population of 3 per cent in the 1975 - 80 period, (this was the actual rate of increase from 1971-1974-75), declining to a 2 per cent rate in the 1995-2000 period.

The projected figure in this table for 1980 seems too high : the World Bank (1980) gives an estimate of 89.1 million as at January 1980, and an annual rate of change of population of 2.7 per cent and this is consistent with the estimated rate of change given in the Second Five Year Plan. However, early figures from the 1981 census give a central population estimate below 90 million. This would imply a rate of change of 2.3 per cent p.a. between 1974 (the last census year) and 1981 - assuming that the censuses are reasonably accurate.

These might seem rather marginal differences, but as we saw earlier in this section, small differences may indicate the beginnings of an important downward fertility shift. According to Eberstadt (op. cit.), the birth rate for Bangladesh is falling, and this seems to be the case, despite the fact that according to Mauldin and Berelson's 'threshold ranges' of socio-economic variables believed to influence population growth (see Table 1.3), none of the values for Bangladesh yet come within the threshold range. But, if a fall in the rate of growth of population is occurring, might it not be due to increasing mortality? Unfortunately, until detailed census figures are available, we cannot test this. However, there is one major piece of evidence that attitudinal changes towards family size are occurring: Maloney et al (1980) in an extremely detailed analysis of the social and cultural bases of population growth in Bangladesh, offer a number of very interesting observations on this subject. Bengali peasant culture, they say historically 'is suffused with a pro-fertility ethos which evolved over 3000 years of adaption and symbiotic relationship between man and land'. However, they found that at the time of their survey, 54 per cent of males, and 60 per cent of females in the child bearing age cohorts now express the preference to have no more children.

We cannot do justice to this large, and complex study here, so we must content ourselves with a few of its general conclusions. Using statistical correlation techniques, the study finds that: the hypothesis of children being desired because of their economic worth (a common claim in the development literature) is _not_ supported; the hypothesis that a large number of children are desired as a sort of replacement insurance against death is _not_ supported; the hypothesis that a reduction in child mortality is a pre-condition of fertility decline is _not_ supported.

High fertility is strongly correlated with religiosity, and with the socio-cultural background which creates the present 'world view' of the rural population. The study recommends an approach to fertility control based on a moral imperative - in effect, to use existing religious and social control systems, which are effective, to reinforce the idea that population control is a moral duty. There is some basis for optimism that this approach will work: 'most people (in the interviewed group) are genuinely apprehensive about the effect of population growth. Most of the village professionals interviewed would like to restrain it'. If this feeling becomes internalized, through teaching at the _para_ (sub-village social unit) level, then fertility reductions will follow.

In recommending this approach, the study specifically rejects the Marxist approach to fertility control of altering production and consumption patterns away from the family, towards the individual basis, on the grounds of it being inappropriate for Bangladesh. It also rejects the idea of State coercion or persuasion, such as has been applied in China and attempted in India, as being unworkable in a situation where the state is neither well known, nor highly trusted. Education, economic and health programmes ultimately _will_ have an effect on birth rates but, the study argues, these cannot be employed as the major instruments now, because from their present extremely low base, they will take too long to become effective (this is an interesting conclusion, in view of the fact that most of the population control effort in Bangladesh at present _is_ based on this approach).

In summary, it seems to us that even if the macroeconomic conditions in Bangladesh are not presently conducive to fertility declines, they may be in the offing anyway. Certainly, there are precedents in other countries for quite significant changes in fertility in advance of across the board economic changes (Sri Lanka comes to mind), and the climate of opinion about the subject seems to be undergoing change in the rural area of Bangladesh.

1.3 Modernization theories and the allocation of development capital

1.3.1 The take-off to development, and the 'Big Push'

In the immediate post-War years, development economists began to seek the key processes which led to development in Western nations, in the hope that this might lead them to some overall economic measures that could be taken in LDC's to accelerate the development process there.

Rostow (1956) and others argued that developed countries have passed through an intensive phase of economic development, after which growth becomes a self-sustaining process. Perhaps the best known theory to result from this work is the claim that something in the order of 10 per cent of GNP in savings investment was the point at which a 'take-off' into high levels of economic growth would occur. Subsequent history has not borne out this figure: many under-developed countries have achieved and maintained savings levels in excess of this figure, with (so far) negligable effects on growth. The complicating factor seems to be one of timing: there are certain phased pre-conditions of development that a country must attain, before it moves from a basic subsistence agrarian state where the government's principal concern will be the establishment of minimal law and order, to a condition where the human, physical and financial requirements for realistic establishment of a modern sector within the economy are in existence (see the discussion of Hagen, below). This is a most important proviso, although it is possible to get the impression when reading some development proposals, and certainly when observing some projects in the field, that it is frequently overlooked. Rostow himself acknowledged that the 'pre-take-off phase' can be very long indeed - and an elementary study of the economic histories of the highly developed countries will provide fairly convincing cases in point.

The accumulation of minimal levels of investment capital then, whilst unarguably a necessary condition of economic growth, can be seen not to constitute a sufficient condition. The capacity of a country to absorb, manage and efficiently distribute capital investment is important, but can tend to be masked by aggregate capital requirement theories. A more specific statement of this argument is to be found in the now fairly well known argument that in most developing countries the limiting factor on the realisation of overall growth is not the availability of money, but the ability to spend it effectively (see, for example, Waterson (1965)). As has already been implied in the Introduction to this book there may exist a third, even more deeply buried set of problems to do with the vacuum that often surrounds planners and aiding agencies: the general unrelatedness of what they do (no matter how capably they may do it) to the problem, aspirations and capacities of the country as a whole.

Much of the post War debate on development centred around industrialization as a key to development. The original statement of the 'Big Push' - a high minimum amount of industrial development more or less to cause a country to 'leap' into a condition of economic development - is to be found in Rosenstein-Rodan (1943). This argument is based on the observation that complementarity between industries exists, and that expansion in one area of the economy requires, for both consumption and input reasons, expansion in others. In addition to reducing the risks of non-sales, this modus operandi will also generate the external economies which accrue to firms within an expanding industry, and to an industry within a growing economy.

Ellis (1958) was among the earliest to criticize the 'Big Push' theorem firstly on the ground that the trade option offers the expanded markets necessary without the need to provide everything through domestic investment. Second, Ellis notes the propensity of 'Big Push' theorists to consider manufacturing as inherently superior to agriculture. Even if this could be generally established, it will become apparent below that a major shift of resources out of agriculture and into the manufacturing sector is certainly not a realistic medium term option for very poor countries, with their predominantly agrarian economies.

In more recent years, the debate has shifted more to this area of the relative merits of the agriculture and industrial sectors. It is fair to say that in the 1950s and 1960s many developing countries pursued industrialization on the basis of 'Big Push' or 'take-off' theories of development. Such policies were put into effect in many countries, and still are in effect to a significant degree simply because considerable capital investment was made and cannot, for social or political reasons, be easily dismantled.

Accelerated industrialization programmes were frequently carried out under conditions of domestic inflation, balance-of-payments crisis and severe pressure on natural resources. Foreign exchange controls were frequently geared to discriminate heavily in favour of the manufacturing sector, often to the extent of allowing an inflow of labour-substituting inputs via overvalued exchange rates. This is the basis of the problem of financial dualism, discussed in a later section of this Chapter.

With the benefit of hindsight, it is now fairly apparent that the results of this approach to development have been disappointing. A superficial glance at the figures may not reveal this: since the War, virtually all LDC's have recorded positive rates of overall economic growth. But there are major qualifications to this. Most importantly, the proportion of the population of LDC's (with some obvious exceptions) below a fixed minimum of nutrition and health has remained the same, or perhaps worsened. Lipton (1977) observes: '.....we have an astonishing contrast: rapid growth and development, yet hardly any impact on the heartland of mass poverty. Among the steel mills and airports, and despite the independent and sometimes freely elected governments, the rural masses are as hungry and as ill-housed as ever'.

It is relatively common to find, in the literature supporting the 'industrialization first' approach to development, arguments suggesting that pronounced inequities are inevitable if rapid early growth is to be obtained, and protagonists are able to cite, if somewhat out of context, Lord Keynes himself on this point(1).

We will return to this distributional issue in the next chapter; for the present, we are interested in the question of whether the industrial sector in LDC's _is_ capable of producing significant growth in aggregate income.

Hagen (1980) has reviewed industrialization in LDC's, and he observes that after thirty years of concentrated effort in pursuit of economic growth: 'per capita product in the LDC's is increasing not only very slowly by the yardsticks of hopes and expectations in these countries but markedly less rapidly than in the industrialized countries'. Hagen, thus, is even less impressed than Lipton with post War development in that he does not accept that <u>overall</u> income growth has been rapid. He cites figures from the World Bank's 1979 <u>World Development Report</u> which show that the average rate of growth in per capita incomes in LDC's since 1960 has been 2.5 per cent, compared to 3.5 per cent in the more developed countries. Hagen's explanation is simple: the marriage of capital intensive high technology production methods with large scale production, which is the essence of growth in the industrial countries, is a complex and difficult system. Attempts to set up parts of it in low-income, non-industrial countries cannot work: there will not be a market large enough to avoid unit costs being absurdly high; the component supply network will not exist, and the alternative of importing will be costly; maintenance costs will be high, and technical expertise difficult or impossible to obtain. The result: 'A capital intensive modern plant in a non-industrial environment soaks up capital that could otherwise be used to the much greater benefit of the country, and operates at a cost that burdens the rest of the economy'.

Hagen provides an identification of the stages of industrialization and he notes that any of them may take a <u>generation</u> to complete:

Stage 1: Self-contained factories - that is, mechanization of production of goods already produced in the country (food processing; leather goods; brick-making and so on).

Stage 2: Initial interrelationships: small casting, forging and welding shops to serve some existing industries; cement manufacture; package making (from <u>imported</u> materials); are the examples cited.

Stage 3: Light engineering - an intensification of capital and complexity in the sorts of activities identified in Stage 2.

The remaining three more advanced stages trace the development of a self-sustaining industrial system through quality control systems, mechanization of agriculture via domestic machine production, technical capacity to operate and maintain complex machinery and, ultimately, appropriate conditions of demand to justify installation of such machinery in interrelated productive processes.

Hagen notes specifically the adverse effects of attempting to short-circuit the above process, via the introduction of heavily protected industry, or industrial systems, too early. The empirical evidence, he suggests, indicates that whilst this approach may hasten progress initially, in the longer term it will tend to inhibit or totally halt it. Of the United Nations listing of least developed countries Hagen says most have not even effectively entered Stage 1. Some of the more advanced African nations are in Stage 2, while areas such as Southern Italy, Colombia and some Central American countries are in Stage 3.

These, then, would appear to be the realities of industrialization in development. No matter what the apparent appeal of mobilizing an available raw material resource may be, they need to be borne in mind.

Because of observations such as these, attention has turned in recent years back towards the agriculture sector - the sector which Ellis suggested should not be considered inferior to the modern sector of an LDC economy. It is reasonable to say that, in pursuit of the industrialization approach, agriculture usually was regarded as an inferior sector, and its (usually) low overall growth rates were frequently cited in support of this view. Agriculture was, in other words, an economic burden that the LDC must bear with until better uses could be made of its mobile productive resources.

1.3.2 Agriculture reinstated

Lipton (1977), in an extensive review of the persistence of poverty in LDC's, argues a very different viewpoint:

> 'The disparity between urban and rural welfare is much greater in poor countries now than it was in rich countries during their early development.... This huge welfare gap is demonstrably inefficient, as well as inequitable..... It persists mainly because less than 20 per cent of investment for development has gone to the agricultural sector..... although over 65 per cent of the people of less-developed countries (LDC's), and over 80 per cent of the really

poor who live on $1 a week each or less, depend for a living on agriculture. The proportion of skilled people who support development - doctors, bankers, engineers - going to rural areas has been lower still; and the rural - urban imbalances have in general been even greater than those between agriculture and industry. Moreover, in most LDC's, governments have taken numerous measures with the unhappy side-effect of accentuating rural-urban disparities: their own allocation of public expenditure and taxation.....; measures raising the price of industrial production relative to farm production, thus encouraging private rural saving to flow into industrial investment because the value of industrial output has been artificially boosted.....; and educational facilities encouraging bright villagers to train in cities for urban jobs ....

Such processes have been extremely inefficient. For instance, the impact on output of $1 of carefully selected investment is two to three times as high in agriculture as elsewhere....., yet public policy and private market power have combined to push domestic savings and foreign aid into non-agricultural uses'.

This is an important argument, not only for sectoral allocations in LDC's in general, but for the case of the forestry sector in particular. This is because forestry can be practiced either as a large scale operation producing raw material for high technology forest products industries, or as a small scale rural based activity, providing basic fuel, food and fodder and structural necessities for rural dwellers. These two extremes will, obviously, exercise very different effects on economic structure, welfare and social factors and so on. This wide variability in feasible practices within the sector provides the basic reason why forestry decision-makers must consider very carefully wider considerations of economic development and philosophy.

1.3.3 A case in point

As in the case of the population matter, Bangladesh provides an informative example of the allocative problem posed by Lipton. To examine it adequately requires a deal of historical and sectoral information: some of this is presented as Appendix A: those readers specifically interested, or not inclined to accept our more general reasoning here, are referred to that area of the book.

Bangladesh exhibits most of the unfortunate economic and socio-economic circumstances which combine into the under-development problem, and it is fair to suggest that it displays extreme manifestations of most of them. Some, such as the major matter of social structure and income distribution, we will return to in section 2.3 of Chapter 2. For the present, we are concerned with the sectoral economic allocations in the country.

The agriculture sector dominates the economy of Bangladesh - it is responsible for about half of the country's GNP, and 75 per cent of total employment. It is based on rice, with jute as the main cash export crop, and important contributions from tea, wheat, sugar and tobacco. Although the land is highly fertile, output is dependent on rather delicate relationships between adequate and timely precipitation, and critical levels of inundation on the deltaic plain which is the predominant feature of the economic geography of the country.

By any standards, the lack of development of the agriculture sector in Bangladesh is a serious problem. Notwithstanding the importance of rice as a basic, staple food, and the good growing conditions for it, yields in Bangladesh are among the lowest of major rice growing countries of the world. Overall, trends in agricultural production make depressing reading.

Table 1.5: TRENDS IN PRODUCTION, IMPORTS AND AVAILABILITY OF MAJOR AGRICULTURAL COMMODITIES IN BANGLADESH

Item	1960-61 - 1969-70	1960-61 - 1976-77
	%	%
Cereals production	2.4	2.6
Fish production	1.8	1.0
Imports of cereals	9.8	7.3
Availability of cereals	1.8	1.6

Source: Table 2, <u>WCAARD Review</u>, Ministry of Agriculture and Forests (1979).

Comparison of the final figure in this table, to rates of change of population will suffice to show the deteriorating basic nutritional status of the nation. Not surprisingly, under these circumstances, it can be calculated (see Appendix A) that real wages in the agriculture sector have fallen to very low levels indeed.

Overall, then, the agriculture sector is characterized by extremely low levels of development and growth. Improvements in technology have been slow, and, as we will see in Chapter 2, where these have been introduced they have if anything exacerbated problems of income maldistribution and impoverishment.

The industry sector, at first glance, appears in somewhat better condition – but as we will see, even this modest achievement has been purchased at considerable cost.

At present, manufacturing industry currently accounts for about 8 per cent of GDP, but some 60 per cent of exports (especially processed jute) come from the sector. About 70 per cent of all manufacturing is owned by large government corporations, and 85 per cent of all industry investment funding is used in these. Many publicly owned industries in Bangladesh continue to incur heavy losses and average capacity utilisation remains low. Within the country, it is fairly common for the capacity utilisation problem to be related to supply factors: raw material constraints; poor maintenance; administrative bottlenecks in spare parts or input importations – and so on. It is at least as arguable (although difficult to prove conclusively) that a fundamental lack of demand may underlie these problems (see the case of the forest products industries, outlined in Chapter 4 below).

At the time of writing, the Military administration of Bangladesh was following policies of return of certain manufacturing businesses to their pre-Independence War private owners, and of sale of other corporation industries to private enterprise. Regardless of the point of view one adopts to this transfer in theory, in practice its effects are likely to be limited: the private capital and investment marked in the country is weak – certainly, it would seem, too weak to take over large, costly and presently highly unprofitable corporation enterprises.

Further evidence is adduced in Appendix A to suggest that, by any standards, the 1970s have not been good years for the area that is now

Bangladesh. The Government itself, in the preamble to the Second Five Year Plan (Planning Commission (1980)) has said:

> 'The common man expected an improved standard of living, a higher material and cultural satisfaction: but this has not happened for a majority of the population. They still live in poverty, in the darkness of illiteracy, and in shanty houses. Over 80 per cent of the population still has income below the amount which is needed to afford them two square meals a day'.

It needs to be said that the 1970s included several natural and political disasters for Bangladesh of a type which, it may be hoped, will not return with such frequency or intensity again. Nevertheless, we will argue below that a good deal of the lack of progress in the country is the result of incorrect perceptions, on the part of sector administrators - and, very often, aid agencies - as to how and where limited development capital should be spent.

To examine this proposition further, we need to consider the stated objectives, and the sector allocations actually made, in past and present economic plans for Bangladesh.

Bangladesh has had three economic plans since its inception in 1970: the First Five Year Plan (FFYP), 1972-77; an interim Two Year Plan (TYP), 1978-80; and the current Second Five Year Plan (SFYP), 1980-85.

The objectives of the FFYP and TYP were stated in rather political, imprecise terms: reconstruction of the war-torn economy; reduction of poverty; creation of employment; and the achievement of social justice. Few would dispute these as overall desiderata for a country in the condition of Bangladesh, but they imply no particular strategy for development, and it is fair to say that for this period, no particular strategy was followed. In retrospect, the overriding aims for the 1970s should have been the achievement of an agricultural surplus, and introduction of effective measures to redistribute at least some of the gains from this toward poorer groups.

What was <u>actually</u> done, in terms of sectoral allocation, can be determined through a review of the capital inputs made over the period. To make a rational comparison, it would seem best to relate development capital inputs to observed real GDP increments. This is done in Table 1.6 by utilizing the following assumptions:

(1) Development capital inputs for a given year are related to incremental output in the following year.

(2) GDP figures for 1979-80 and 1980-81 are projected on the basis of the 1972-73 - 1978-79 trends for each sector.

(3) GDP increment is defined here as the accumulated excesses in years 1973-74 - 1980-81 over the base figure for 1972-73. Sectoral breakdowns are calculated on the same basis.

It is apparent that these figures do not capture all of private investment. In agriculture, particularly, it would be impossible in Bangladesh to calculate (or indeed even to classify) the proportion of farm income recycled as investment. It is more likely, however, that the figures given above represent most of expenditure on major _additions_ to the productive base in the respective sectors.

If this contention is accepted, the figures reveal some interesting features of sector allocation. Most obviously, it can be seen that agriculture provided some 48 per cent of incremental GDP, but received less than 30 per cent of government development capital, and even less of total development capital inputs. The 'total investment' capital/output ratios given are analogous to those used by Lipton in his 'k - criterion', (within the definitional limits already discussed). It should be noted that agriculture almost managed to maintain its contribution to incremental GDP over the period, despite the relatively low development capital inputs to the sector, which suggests that in this case (as for Lipton's general findings for LDC's) the marginal efficiency of capital in agriculture is considerably higher than it is elsewhere in the economy.

There are, of course, a number of arguments against a purely numerical approach to the allocation of capital amongst sectors:

Table 1.6: GDP SHARES AND SECTOR ALLOCATION 1972-73 - 1980-81 (1972-73 PRICES) (b)

Sector	1972-73 (Tk10⁷)	1980-81 (Tk10⁷)	% change	Total increments to GDP 1973-74-1980-81 (Tk10⁷)	Total allocation government development capital 1972-73-1979-80 (Tk10⁷)	Total(c) private capital investment 1972-73-1979-80 (Tk10⁷)	Capital/Output ratios Total investment	Capital/Output ratios Public investment
Agriculture	2 883	3 796	3.5	4 075	1 460	94	0.38	0.36
Industry	520	596	1.7	331	722	248	2.93	2.18
Power	236	314	3.7	188	304	(na)	-	1.62
Housing	15	69	20.8	342	718	375	3.19	2.10
Total GDP	5 005(a)	7 009(a)	4.3	8 571(a)	4 810(a)	978(a)	0.67	0.56

(a) Includes trade, transport and social services not shown in the sector allocations. (b) Deflated from SFYP current values using agriculture industry price indices for 1972-73 - 1977-78, and official overall inflation rates for 1978-79 and 1979-80. (c) The private investment figures given here relate to expenditure on productive capital items. They do not include 'non-monetized' investment - such as mass mobilization of voluntary rural labour for infrastructure (see discussion of this in SFYP context near Table 1.7 below).

Sources: Tables 1.2 and 9.2, Appendix 1, SFYP.

complementarities between industry and agriculture; the difficulty of comparing industries at differing states of development or establishment; structural rigidities; absorptive capacity constraints; and so on. These arguments have been widely discussed in the literature, and this is not the venue for a reiteration of cases for and against them. Elements of all of them probably do apply in this case, but there is a question of degree involved. One would need to assume very rapidly diminishing returns to additional inputs into agriculture, or equally significant (and imminent) expansions in productivity in the non-agriculture sector, to justify the actual pattern of distribution on economic grounds. The latter possibility seems remote in Bangladesh, particularly when taking into account the fact that the industry sector there is by no means new, and most manufacturing plant is anything but recent. And it seems equally unlikely that, starting from such a low present base of development capital, the efficiency of use of capital in agriculture could drop off so rapidly, even if quite significant additional amounts of it were to be made available.

On the bases of the above data and assumptions, our conclusion is that, whether or not an urban bias was intended in allocations made under the FFYP and the TYP, it seems to have been the result.

In view of this, an important aspect to examine in the current Second Five Year Plan is the interpretation it makes of past events. This will allow some deduction of what lessons have been learnt by planners. If, for example, it were to be found that past failures or inadequacies were attributed largely to natural causes, adverse exogenous market and trade developments, and so on, then one could fairly safely assume that the approach and strategy shortcomings in past plans would be repeated in the current one.

The first thing to note, in considering this, is that as exemplified in the quote from the SFYP given earlier, it is made clear that recent economic progress in Bangladesh has been disappointing. Whilst the document does identify a number of exogenous causes for this (weather conditions; the war and subsequent loss of Pakistan markets; international trade and commodity difficulties, with resultant inflation and recession; and so on) it also notes a number of much more politically 'live' problems. Government intervention in the rural sector is criticized for being inefficient and ineffective; industrial unrest and chronic over-manning are cited as major causes of low industrial productivity; a tendency to undertake more projects than can properly be managed is noted and programming weaknesses are said to be severe; the Planning Commission itself (the source Department of the SFYP) is identified as a major point of failure in the system; an over-commitment to international aid agencies instead to the development programme itself, is also cited as a problem. By and large, the authors of the SFYP make what seems an honest attempt to specify previous mistakes or inadequacies, and in the process they have not refrained from identifying the culpability of the Government itself.

The next major matter to consider is whether or not the national objectives and strategy in the SFYP give adequate direction to sector administrators, and whether that direction is consistent with what seems to be the generally accepted view of priorities for development of such an economy.

The overall priority which is being strongly advocated in Bangladesh by the Government is achievement of self sufficiency in food production. At the highest levels of planning, all other developments are seen as being linked to this. For example, in an address to the National Economic Council, the late President said:

> 'Our economy cannot be stable unless we increase agricultural production and become self-sufficient in food. So our entire planning in industry should be organized to help agriculture in increasing its production.'

In the SFYP document itself, the objectives given are a mixture of the usual general statements of good intent about increasing the rate of economic growth, reducing unemployment and so on, with more definitive statements on population growth, agricultural production, mass participation in village level governing bodies and literacy programmes.

In discussing food production aims, the SFYP makes specific mention of the landlessness problem (to which we will return in Chapter 2): in response, it seems, to the expressed concern of the Government on this point. It stops short of radical land ownership reform (and has already been trenchantly criticized for doing so)(2), proposing instead the introduction of fiscal and legal measures to encourage land management at the village, rather than the household level. The Government sponsored programmes of <u>Swanivar</u> and <u>Gram Sarkar</u> discussed earlier are two such schemes. Whether they will succeed is a moot point, but in our view, the present Government is undoubtedly correct in its assessment that it <u>could</u> not introduce land reform at this time.

So far as the industrial programme is concerned, the SFYP reiterates the priority for industry to support, rather than compete with, agriculture. This is qualified somewhat by reference to the need for structural change towards more industrialization but, as expressed in the document, this is a policy for the future, rather than the present.

Overall then, if we read no further in the SFYP than this, the most reasonable interpretation of it would be: that it advocates a strong rural development basis (the term 'rural bias' is in fact used in the document itself); that it at least displays an awareness of the structural and market problems inherent in forced industrialization programmes; and that it recognizes that something must be done to involve landless groups in economic activity, so that whatever gains made will tend to be self-sustaining.

However, an examination of the allocation of development expenditure amongst sectors in the SFYP creates a rather different impression. It is possible to repeat the exercise described at Table 1.6 above for the projected outputs and expenditures. The assumptions involved are more or less the same. The results are given in Table 1.7.

There are some provisos to account for when interpreting these figures. As in the case for FFYP and TYP data, the private investment figure is open to question. In this case, the SFYP provides an estimate of projected 'non-monetized' investment in the various voluntary mass mobilization schemes to create infrastructure and so on, but this is

Table 1.7: VALUE ADDED AND SECTOR ALLOCATION 1979-80-1985-86 (1979-80 prices)

Sector	1979-80 GDP (Tk10^7)	1985-86 GDP (Tk10^7)	% change p.a.	Total increments to GDP 1980-81-1985-86 (Tk10^7)	Total financial outlay 1979-80-1984-85 (Tk10^7)	Public sector financial outlay 1979-80-1984-85 (Tk10^7)	Capital/Output Total	Capital/Output Public
Agriculture	88 860	128 347	6.3	130 714	74 350	65 000	0.57	0.50
Industry	11 911	31 049	8.6	39 493	48 850	32 750	1.11	0.83
Power	3 793	8 980	16.0	16 743	29 150	29 150	1.74	1.74
Housing	13 670	17 498	4.2	12 932	22 200	12 200	1.72	0.94
Transport	8 845	15 628	10.7	22 873	37 200	36 250	1.63	1.58
Total	179 769(a)	272 832(a)	7.2	306 827	255 950	201 250	0.83	0.66

(a) Includes communications, health, education, welfare plus (possibly) some amount for construction not shown in the sector allocations.

Sources: Tables 3.2, 3.5 SFYP.

excluded from the above figures on the grounds of its uncertainty. Were it to be included as part of development capital at the value estimated in the SFYP, it would raise the total capital/output ratio for agriculture to 0.65, and that for housing to 2.27, whilst other sectors would alter only marginally.

Secondly, these figures are, of course, projections, and they may be quite different to what actually eventuates. Indeed, in the time which has elapsed since their publication, there are already some indications that levels of funding available will be less than those given above. However, the interest here in these figure is in what they reveal about the overall intentions of the SFYP.

Overall, development capital expenditure on agriculture in Table 1.7 actually falls slightly, to 29 per cent of total allocations, although the Government figure rises slightly, to 32 per cent. The allocation/output ratio for this sector is again considerably lower than for the others. There is, certainly, a deal more in real terms projected for allocation to the sector, and its projected contribution to incremental value added is 43 per cent, which is less than the 48 per cent figure for previous years. However, this may be due to what seem to be extremely optimistic projected growth rates in some of the non-agricultural sectors.

The same qualifications to interpretation of the ratio figures in Table 1.6 apply again here. However, the overall impression they create remains unchanged: the sectoral allocations under the SFYP remain biased towards expenditure in the urban, industrial sector of the economy, despite the stated intentions in the document. Industrial refurbishment and development programmes, power generation projects, urban housing and so on seem, once again, to have claimed larger pro rata shares of capital than the sector which, ultimately, must support them all if whatever gains made are to become self-sustaining. Under these circumstances there is a danger that health and education programmes, which at the very least ought to remain value neutral in terms of urban/rural distribution, will also be disproportionately drawn into the expenditure centres.

It is worth attempting some analysis, or at least reasoned speculation, on the causes of this persistence of perverse capital allocation in economic planning in Bangladesh. There are, of course, a very large number of possible reasons for it, and it may be that more than one actually applies. However, there are a few obvious possibilities which are worth closer examination.

To the more cynical observer of the development scene, perhaps the most obvious explanation of all is that the Government wishes, for international political purposes, to appear to be supporting rural development, but in fact perceives its own internal interests, and those of its strongest urban and rural supporters, as being based on continuation of the <u>status quo</u>. This explanation implies a unified conspiracy between politicians, upper level planners and sector administrators to create a plan which, crudely speaking, is intended to say one thing but do another. A somewhat less sinister variant of this explanation would be that the Government genuinely believes in the industrialization approach to development, but prefers not to be publicly associated with it.

These explanations are, in our view, far too simplistic to be of any use. It is certainly true that no government which wished to survive in Bangladesh could afford to ignore recent history to the extent of disregarding the aspirations of the highly volatile urban population. But this is not the same as suggesting a conspiracy between Government and sectional interests, to keep the poor poor, as it were. It would be more logical to view it as a political constraint which, roughly translated into economic terms, might have required the Government to <u>maintain</u> allocations directed at the urban population in real terms. Given the expanded amount of development capital available, this would have allowed the agriculture allocation to move up to something like 40 per cent of the total. This may not have suited some development theorists, but, had it materialized, it would have been a considerable achievement. Whilst it is always difficult for an outsider to judge political motives, it seems to us that an allocation based on maintenance of urban/industrial programmes and an expanded rural programme would have found considerable support amongst elected officials and senior economic planners.

A second possible reason for the allocation which has resulted is the presence in Bangladesh of a very large number of bi-lateral,, multi-lateral and voluntary development assistance agencies. At the present time, about 75 per cent of Government development capital expenditure is externally financed. Many of the agencies present represent specific political, development or even commercial interests, and can be expected to attempt to influence planners and policy makers towards a particular view of development. If they are persuasive enough, and if there is inadequate central control and vetting of project and programme proposals, then the resulting plan will be an unco-ordinated collection of special interest projects.

Clearly, the influence of donor agencies has been a matter for concern in Bangladesh: it is remarked upon in the SFYP document, and it is also mentioned in at least one important work about planning in Bangladesh (see Islam (1979)). However, as noted above this sort of problem can only become and remain serious if the Government is not, for one reason or another, controlling and co-ordinating the activities of outside agencies around a central economic strategy.

This leads us to the third possible cause for the dichotomy between strategy and allocations in the SFYP: it being that a severe communication gap exists between the Government with its upper level policy advisers and planners on the one hand, and the longer established and more bureaucratic sector administrations on the other. The plan document resulting from such a system will be unable to reconcile the objectives, priorities and national strategy, with the more or less traditional sector programmes which are submitted. It is important to recognise, in this context, that unless calculations of the type implied in Tables 1.6 and 1.7 are done, the relative capital-productivities of the various sectors may not become apparent. These sorts of calculation are not made in the plan document.

Under such circumstances, it does seem at least feasible that a breakdown in communications about the nature and intent of overall national objectives, and the role of sectoral programmes in achieving these, has occurred. The overall design and strategy for the SFYP was carried out by the political wing of Government, with strong inputs from the central economic evaluation area of the Planning Commission, and some from outside academics. The resulting design and objectives were communicated to the sectoral areas of the Planning Commission which, in consultation with the relevant line Departments, formulated sectoral programmes.

In the situation where the central economic unit is unable to monitor and adjust each sectoral programme individually, such a system is likely to produce sectoral programmes fairly similar to past ones, given the continuity of personnel and policies in the established Departments (many of which, in Bangladesh, have existed in one form or another for quite considerable periods of time). Added to this is the problem that in sectoral areas of the Planning Commission itself, a situation seems to have emerged where economic qualifications seem to be very much secondary to specific sectoral knowledge. In practice, what seems to have occurred is that the central economic division of the Planning Commission was swamped by the sheer technical detail and volume of the sectoral programmes, and was unable to respond adequately with close analyses of whether the projects and programmes included were as consistent with overall objectives as possible.

There is, as we have seen, a strong basis in the current literature on development for very poor, agrarian countries to emphasize rural productivity and income redistribution in their development programmes. At the highest political and planning levels in Bangladesh, this message has - at least to some extent - been received, and understood. The SFYP proposes a reasonable and realistic strategy, and displays commendable honesty in specifying the mistakes that have been made in the past.

However, it seems likely that the steps toward development implicit in the SFYP strategy will only be taken very slowly, if at all. This is not because the sector administrations are incapable of carrying out the necessary measures, but rather because they seem to lack adequate understanding or conviction of them to do so. Although this is a rather different problem to the commonly cited ones of implementation or absortive capacity constraints, a good deal of the aid programme in Bangladesh still seems based on the idea that further inputs of technical expertise and equipment are what is required.

Chapter 2

INCOME DISTRIBUTION : THE CRITICAL ISSUE

Although it may not always be specifically recognised as such, the income distribution question lies beneath virtually all of the major issues in development economics at the present time: decisions as to <u>what</u> and <u>how</u> an LDC enters the international trade market are important determinants of the structure of the domestic economy, and how it allocates its rewards. The international market is itself the major means of redistribution of income between nations (in combination with - or perhaps opposition to - the transfers of international aid). Allocation of development capital between sectors in the economy has implications not only for overall economic performance, as we saw in the previous chapter, but also for overall income equity and the ability of the economy to generate <u>and</u> <u>maintain</u> effective demand. And, within individual sectors - particularly the large agricultural sector that typifies most LDC's - linkages between aggregate sectoral development policies, and growing landlessness and impoverishment that ultimately must threaten the overall growth dynamic, can no longer be ignored.

2.1 Trade, tariffs and the new international economic order

There has been a deal of controversy in the development economic literature, and in the political and administrative structures of many LDC's themselves, on the trade option for development. Frequently, the debate is couched in terms of nationalism, self-reliance and so on, rather than in economic terms. However, it is fairly apparent that the question is one of degree, rather than of the absolutes in which it may be stated. It may be possible for very large developing countries, such as India and China, to opt for extremely low levels of international trade, but for most developing countries there is an insufficient resource base, and usually too little depth and diversity in the economy, to allow such an option to be pursued for very long.

In the earliest phases of development, trade is linked to the spread of the money economy. Even when it is possible for the rural sector to produce (tradeable) surpluses, these will only eventuate when some inducement to trade is offered. In the absence of such inducement, the subsistence economy, by which is meant here a non-money economy rather than a necessarily survival situation, will remain intact.

Ultimately, if the standard model of economic development applies, some degree of specialization by rural producers will begin to appear, and money as a medium of exchange will become more significant. Once this pattern emerges, producers will be better able to utilize the market process, but by the same token they will be more vulnerable to fluctuations in it. It is relatively unusual, when a trading pattern emerges, for it to be reversed. Irreversibilities, based on what is sometimes known as the 'revolution of rising expectations' and other factors, seem quite strong.

As will be well known to most readers, the classical theory of international trade holds that free trade between nations permits specialization, and expansion of the total consumption possibility. Moreover, the theory of comparative advantage shows that a nation need not have an absolute advantage in production of any commodity, compared to another country, for there to be mutual advantages to both in the trade option.

The classical theory admits only two important exceptions to the generally beneficial effects on consumption and welfare of the free trade condition: the optimum - tariff argument, and the infant industry argument. The first is based on the possibility that a country with monopoly or monopsomy power can use a tariff to restrict trade but alter the terms of trade in its favour: this is an interesting possibility, but not one which is of primary concern for the sort of countries we are dealing with here. Some cogent discussion of it can be found in Johnson (1965). The second exception, the infant-industry argument, is certainly one which is relevant to many developing countries, which use tariff or non-tariff protection specifically in an attempt to shield domestic industries from import competition until, it is hoped, they become mature and efficient. Economists have raised three basic difficulties with the infant-industry argument. First, it can be shown that infant industry possibilities might be better in the export sector in many countries, than in the import competing sector - in which case a high tariff structure will damage overall prospects by raising input costs. Second, it is preferable from both the efficiency point of view, and also in terms of political and commercial accountability, for direct, visible subsidies to be used to promote desired industry development. Third, it has been frequently argued that, in effect, industries sheltered as infants never grow up: the protection becomes institutionalized and permanent. Johnson (1965) has shown that in fact this argument is inconclusive because it is quite possible for a nation to make welfare gains from a protected industry, even if that protection continues in perpetuity: the situation produced cannot be optimal, but it _can_ be better than the situation which prevailed prior to establishment of the industry in question. Nevertheless, Johnson emphasizes that modern theory does _not_ support the use of tariffs on the infant industry argument: in any case (except the optimum tariff one) where tariffs can be employed, subsidies can be applied with greater efficiency, and hence with greater positive effect on overall welfare.

We are concerned, in this review, with the implications that recent developments in world trade have had for LDC's, and with some of the remedies or approaches that have been proposed.

The Bretton Woods Conference in the 1940s, established, among other things, the principle that national commercial policies should be subordinated to international interests. The creation of the General Agreement on Tariffs and Trade in 1947 was designed to foster a freer trading approach, albeit that quotas, exchange controls and the infant industry argument were all tolerated under the agreement.

Under these arrangements, developed countries have moved towards trading arrangements between themselves, and this has allowed LDC's to participate in trade in manufactures to a larger extent than was previously the case. Whilst this has benefitted LDC's capable of supplying such goods in any quantity, it excludes a significant proportion of the under-developed world. Moreover, in recent years, as protectionism in general has risen in response to prolonged recession in the West, many developed countries have imposed 'voluntary' quotas on imports of high volume manufactures from LDC's.

In agriculture, the situation is perhaps even less favourable: such countries as the United States have long drawn a distinction between trade in manufactures and that in agriculture, effectively excluding the latter from international agreements and understandings. Moreover, the formation of the European Economic Community (permitted under the exemptions in GATT for formation of free trade areas and customs unions), has permitted the subsequent formation of the European Common Agricultural Policy. There is no doubt that the effect of this policy has been to discriminate against agricultural exports from non-European countries. Much of this effect has impacted on advanced agricultural export countries, such as Australia, Canada and New Zealand, but it has also had a disruptive effect on Latin American countries, and also on certain African countries which have not found favour with the policy-makers of the European market.

Many LDC's, by nature, rely on the export of a few primary commodities for their foreign exchange, and as such they will remain subject to the discriminatory practices on agricultural trade - whether in technical violation of GATT or not - for as long as large individual developed countries, or groupings of countries, see it in their interests to persist with such practices. Demand for food and other basic primary products does not move upwards proportionally to income, although this, as Myint (1973) and Meier (1976) point out, is not an argument in itself for abandoning the concept of primary exports : quite appreciable <u>absolute</u> expansions in exports remain possible. There is, in fact, a great variability in the fortunes of countries which have relied substantially on primary exports : Australia, for example, has on balance done reasonably well out of wool. On the other hand, Australia's fruit growing, New Zealand's dairying and Ghana's cocoa production are examples of industries which have been severely affected by restrictive agricultural trade policies.

It is occasionally tempting for LDC's to theorize about utilizing their supposed joint-monopoly power over tropical products: coffee, cocoa, tea, sugar, rubber - and the range of mineral products they possess - tin, bauxite and copper, for example. There have even been suggestions that wood may be in this category. The assertion of market power by the OPEC cartel in the early 1970s may have given further impetus to such ideas - but the essential uniqueness of oil, with its fairly restricted occurrence (in large volumes), and its fundamental supporting role in the whole massif of Western industry, should be given considerable weight in this context. Substitutes exist for many tropical LDC primary commodities, and in any event, where the buyers tend to be better organized than the sellers, the prospects of selling cartels working for long is remote. It is no accident that rubber production has shifted substantially from Latin America to South East Asia, for example: it took only some fairly rudimentary attempts to exploit an existing monopoly supply situation for this to happen.

None of this, of course, is to suggest that LDC's should not attempt to look after their own interests by cartelization, supply restriction and so on, if there *are* prospects that these will work. But we are less optimistic than some that much potential for the unadorned market warfare approach exists for LDC's, given that their adversaries - if we persist with this militaristic analogy - are themselves in control of some vital goods, and are by no means disinclined to use their market power in their own interests.

It may be for this reason that a great deal of interest has been displayed in recent years in two alternative approaches to the trade question:

(a) renegotiation of the international trading relationships, in such a way as to guarantee LDC's reasonable access for their products on major markets;

(b) a total re-think of the whole attitude of LDC's to trade; in effect, a withdrawal from the 'high trade' to a lower trade option.

The latter development has arisen largely out of Dependency Theory, which we will discuss under the general income distribution head later in this Chapter, since it is based on redistributional philosophies.

The former topic leads to a discussion of the New International Economic Order and the associated matter of the North-South dialogue.

At the risk of over-simplification, the NIEO has two basic thrusts. Firstly, it seeks to remove randomness and political considerations from aid-funding. Aid, it is argued, should be regarded on more or less the same basis as welfare assistance within the more developed countries: allocations should be based on some minimum needs assessment, made as a matter of right - probably through the operation of some sort of international taxation system based on overall GNP, or defence expenditure, or some other factor. Secondly, the NIEO seeks to rationalize the system of international trade, specifically to remove impediments to exports from developing to developed areas of the world. As part of this, it is generally argued that regulation of the exploitive or otherwise counter-productive behaviour of multinational corporations operating in LDC's is required.

We have already discussed the matter of trade discrimination, but in considering the NIEO, we need to look a little more closely at its nature and extent, in order to come to some reasoned judgement as to what LDC's might expect from continued negotiations in this area.

Fried (1975) provides some figures to show that in the fifteen years to 1975, exports of manufactures from non-oil exporting developing countries (NOEDC's) have in fact expanded quite rapidly - at a rate of some 12 per cent per annum - such that by 1975, these accounted for 40 per cent of all exports from the developing countries. However, there are two points to bear in mind here: firstly, these figures reveal nothing about the existence or effectiveness of trade barriers, since they provide no basis for comparison. It has been suggested by Robert McNamara(3) that removal of trade barriers by the developed, Centre countries could transfer something in the order of $30 billion (at 1975 prices) in foreign exchange earnings to the developing countries in ten years.

The second point to bear in mind is the one raised earlier: the NOEDC's are by no means an homogenous, equal group. Indeed, it is possible that a very significant share of the export earnings for manufactures is accruing in relatively few countries at the upper income end of the developing country scale: the so-called Newly Industrializing

Countries. Some of these countries may in fact be due to leave the ranks of the NOEDC's, and it is of passing interest to note that the Far Eastern Economic Review, in its 1979 Yearbook, identifies a tendency for Western countries to seek to involve some of these more advanced developing economics in GATT - the General Agreement of Tariffs and Trade - in effect, to split them away from Third World associations such as the Group of 77.

So far as the overall situation of NOEDC's is concerned, Bhattacharya (1977) has provided some useful figures on this subject, and his analysis is highly relevant.

Basically, Bhattacharya argues that NOEDC's have been heavily exploited through trade with the centre countries, and he bases his case on trade figures since World War II. Table 2.1 below extracts figures from the various tabulations given by Bhattacharya.

Table 2.1: NOEDC VALUE OF EXPORTS AND IMPORTS IN CURRENT US$ (X 10^9)

Year	f.o.b. exports	c.i.f. imports	Balance	Investment income payment	Other items	Current account balance
1950	14.8	14.5	+0.3			
1955	17.0	19.4	-2.4			
1960	18.9	23.7	-4.8			
1965	24.3	30.1	-5.8			
1970	35.8	44.4	-8.6			
1971	36.2	49.2	-13.0			
1972	42.3	53.6	-11.3			
1973	61.2	73.3	-12.1	-6.1	6.9	-11.3
1974	85.5	115.8	-30.3	-7.5	8.1	-29.7
1975	83.9	123.0	-39.1	-9.7	6.1	-42.7
1976	97.2	126.1	-28.9	-13.0	8.8	-33.1

Source: Bhattacharya (1977), Tables 1 and 7, basic data UNCTAD 1976 Handbook of International Trade and Development Statistics Tables 1.2, 1.4; UNCTAD TD/B/642/ Add 1 (1977); UNCTAD TD/186.

Bhattacharya cites work by Balassa, and Little et al., to support his claim that trade and tariff policies adopted by the Centre countries have discriminated against NOEDC's to a greater extent than those levied against other Centre countries. He identifies the activities of trans-national corporations, the prevalence of 'tied' aid, and the operation of the international monetary system as contributing causes to the current imbalances.

In 1976, the value of trade between NOEDC's and Centre countries was $141 billion. Bhattacharya suggests from the historical data that terms of trade have deteriorated 20 per cent since 1954, and thus he reasons that the trade loss figure in 1976 can be put at $28.2 billion. To this, he adds investment income payments of $13 billion, and a 'brain drain' factor which he estimates as $5 billion. Thus he calculates a total of $46.2 billion has been 'lost' by NOEDC's to the Centre. This figure, when offset against financial flows of $39.2 billion in the opposite direction, gives an annual net loss figure for NOEDC's of $7 billion.

Bhattacharya refers to the accumulation of $102 billion in reserves between 1970-74, and he suggests that price rises for manufactures sold by the Centre to NOEDC's in 1975 were not due to OPEC oil price rises, since these were not substantial in that year. Bhattacharya makes the point that his total loss figure does not include tariff losses.

There are a couple of points of interpretation that we should consider at this stage:

1. Despite Bhattacharya's argument on the 1975 figures, it does seem likely that a good part of the surpluses he identifies have in fact accrued in OPEC countries. The OECD Half Yearly Outlook for 1979 shows that the surpluses occurring to OPEC are currently in the order of $30 billion per annum. OECD estimates that the indirect effect of oil price rises might be 2 to 3 times the direct effect. It seems very likely that at least some of this indirect effect might lag considerably behind direct effects, and this makes Bhattacharya's argument for 1975 difficult to accept. It is, of course, of little consolation to NOEDC's that a significant part of import price rises in recent years may have originated outside the Centre, but nevertheless this would seem an important point to bear in mind when considering policies and options for international redistribution under the New International Economic Order. It is one thing to expect Centre countries to redress long term trade

imbalances between themselves and NOEDC's. It is quite another to ask them to compensate NOEDC's for losses to the OPEC group - particularly in a situation where Centre countries themselves have sustained losses to the same source.

2. There is no reason why any further allowance should be made in the trade loss figures for tariff payments, although Bhattacharya seems to imply there should be. Tariffs, in this context, are paid by consumers in the Centre countries, not by NOEDC's. If Bhattacharya intends to argue that the cost of trade losses due to tariff impositions is not included, then this also is not sustainable, because the measured trade figures must have implicit in them any trade effect from tariffs.

It is interesting to compare import: export ratios from the NOEDC viewpoint. From Bhattachaya's figures, this ratio demonstrates no particular trend for the 1970s, despite the very large oil price induced effects. Some more specific figures for the Asian region can be extracted from the Far Eastern Economic Review Yearbook for 1979. These are summarized in Table 2.2 below.

Table 2.2: ASIA TRADE FIGURES 1975-77 ($ million)

	Imports c.i.f.			Exports f.o.b.		
	1975	1976	1977	1975	1976	1977
Bangladesh	1 267	865	1 174	327	401	448
India	6 135	5 094	6 355	4 299	5 013	6 637
Thailand	3 280	3 573	4 711	2 208	2 980	3 565
Hong Kong	6 767	8 882	10 466	6 019	8 526	9 616
Malaysia	3 599	3 925	4 985	3 831	5 295	6 153
Taiwan	5 960	7 609	8 511	6 302	8 156	9 361

Source: Asia 1979 Yearbook: Far Eastern Economic Review.

For present purposes, the figures have been split into two groups: the less developed and the more developed of these Asian countries. They show an expected result, in that for the less developed group, import : export ratios exceed unity. Whilst caution is needed in interpreting only three years' data, it is interesting to note that the import: export ratio for the less developed group fell substantially (by

26 per cent) over the period. This was achieved by substantial export gains, combined with only modest increases in imports. By contrast, imports rose by 32 per cent for the more developed group, largely (presumably) as a result of the oil price effect in those years.

The overall trade position of NOEDC's is not, on the basis of the foregoing discussion, as clear-cut as some of the stronger protagonists of the New International Economic Order might suggest. Undoubtedly, as argued at the beginning of this Section, there *is* discrimination present in the policies of the Centre countries. It is equally apparent, however, that whether by accident or design the Centre countries are not recouping all of the trading losses imposed by oil price rises through the operation of these policies. In fact, the net losses currently being incurred by NOEDC's on either Bhattacharya's or McNamara's calculation, are rather low (4). Adjustment to a value-neutral trading policy between NOEDC's and the Centre therefore, whilst perhaps morally satisfying, might not represent a significant gain for NOEDC's. Yet, to be realistic, it is probably the best outcome they could hope for from the New International Economic Order.

Whether even this fairly modest adjustment occurs will depend to some extent on the economic ideologies being followed in Centre countries. A useful means of considering this matter is to briefly review some of the tenets, and the surrounding debate, of the Brandt report (Brandt et al 1981). Basically, the Brandt report argues that stimulation of the NOEDC economies by increased aid flows from Centre countries will, ultimately, be in the interests of Centre countries, through the beneficial effect of resulting increased trade on their presently depressed output levels. This is, in effect, a Keynesian prescription for curing economic recession, in an international context.

Supporters of this approach (see Jenkins (1979), for example) argue that in terms of the world economy, Keynesian prescriptions of demand stimulation are still relevant, even where they may not be for individual countries. Critics of Keynesian policies, for example, point out that the major impact of expansionary fiscal policies will be felt in adverse balance of payments developments: but in the closed, world economy, this is not a problem. Moreover, in a closed system, the size of the

Keynesian multiplier will be larger than in an open one, where leakages occur.

Monetarist economists criticize the Brandt report for paying insufficient attention to what they see as the real causes of econoic recession in the Centre countries : an over-expansion in the world's money supply, leading to inflation and therefore the need for restrictive, deflationary policies. Brandt argues that a deficiency in world aggregate demand is the basis of the problem - but the monetarists claim that expansionist policies as recommended in Brandt will, at best, stimulate output only in the short run, and will lead inevitably to higher rates of inflation in the longer term.

Henderson (1980) emphasizes what seems to be an internal inconsistency in the Brandt report. The central argument of the report is, as we have seen, that large scale transfer of resources from Centre to NOED countries will make a major positive impact on growth in both. But elsewhere in the report, the question is raised : why <u>should</u> Centre governments, following restrictive fiscal policies within their borders, attempt simultaneously to stimulate, albeit by the indirect means of recycling through NOEDC's? The reason, the report claims, is that higher exports to NOEDC's would be balanced by higher imports from them - so no serious inflationary effect will result. But Henderson points out that this is inconsistent with the Keynesian approach advocated elsewhere: a balanced trade expansion of this type could only have a very minor effect on output and employment in Centre countries: gains in exporting industries would be offset by losses in import competing ones. In a static mode, Henderson's observation would be correct: in a dynamic model, which postulated some net gains to comparative advantage in trade, and some real increases in demand to offset import - competing industry contraction, his argument is less forceful. Once again, the argument reduces to the view taken of Keynesian prescriptions in this situation.

It is fair to say that monetarist theories predominate in economic policy - making in Centre countries at the time of writing. As exemplified above, monetarist economists have not been slow to suggest that the inadequacies of expansionist policies at the national level will be no less severe in the context of international Keynesianism. Whilst these ideas hold sway, there would seem to be little possibility of widespread political acceptance of the Brandt recommendations in Centre countries. Progress towards a New International Economic Order is, therefore, likely to remain very slow indeed.

2.2 Financial dualism, the monetary system and disguised unemployment

Related to both the trade question and industrialization, is the system of exchange which develops in a country emerging from a purely subsistence economy. In the subsistence economy, each family carries its own 'subsistence fund' of food; savings are additions to this stock, and capital is extra land or means by which to work it.

When a country deliberately opts for a policy of rapid industrialization, it is common for this to be approached via 'easy money' policies: low interest rates and abundant credit for the modern sector of the economy. However, in the rural sector, credit will remain scarce, and interest rates high, and even usurious. In fact, the credit problem for small farmers can be worsened by a siphoning of government and institutional credit into the modern sector. This situation constitutes the well known phenomenon of financial dualism in developing countries, and it is very much a distributional, as well as efficiency issue.

Since attaining political independence, many developing countries have also attained a degree of monetary independence, at least to an extent which allows them to pursue dualistic and other policies. In theory, there are some advantages to this: the country can pursue counter-cyclical policies to ameliorate rapid exchange rate fluctuations; the government has the option of varying the money supply independently of exchange holdings, and can undertake deficit financing if it so chooses. On the other hand, Myint (op. cit.) has argued that advantages such as these are often outweighed by the 'practical consequences of the freedom to pursue inappropriate fiscal and monetary policies' - an observation which obviously need not be restricted to developing countries. But monetary instability, and unmanaged dualism can have disastrous effects on the (usually fragile) investment capabilities of the small scale private sector in developing countries.

As noted above, one of the major justifications for adopting dualistic strategies in developing countries is the belief in the necessity to promote the modern sector of the economy through industrialization, and an important factor in this is the existence of 'disguised unemployment' in the rural sector: the perceived need to utilize more fully the under-occupied human resources in the subsistence based rural sector. However, whilst the disguised unemployment theorem suggests that removal of surplus labour from the rural sector will not affect production, the more convincing argument seems to be that where the subsistence economy persists, and therefore there is an absence of incentives to produce a surplus, it is more likely that food output will fall. Thus, Myint and other critics of the industrialization model, have pointed out that where land is the scarce factor, it is in fact economically logical within the subsistence framework to utilize labour to the point where its marginal productivity is zero. Lewis (1963), in presenting what is generally taken as the case for utilization of under-employed rural labour elsewhere in the economy, does not neglect to identify the need for parallel developments in agriculture if industrial development is to be sustained: 'The moral is simply that measures to increase the productivity of manufacturing industry (whether cottage or factory) must be paralled by measures to increase the demand for manufactured product..... If capital is being put into developing manufacturing industry while a country's agriculture remains stagnant, the result is bound to be distress in the manufacturing sector, as factory and cottage workers compete for a limited demand'. Indeed, the history of this argument is as old as the formal study of economics: two hundred years ago, Adam Smith proposed, in The Wealth of Nations, that free trade, and higher production of agricultural wage goods, were the basic ingredient of sustainable industrial growth. Thus: 'it is the surplus produce of the country only, or what is over and above the maintenance of the cultivators that constitutes the subsistence of the town, which can therefore increase only with the increase of the surplus produce.'

2.3 Development and income distribution

As noted at the outset of this chapter, the central theme of the development debate has become the income or welfare distribution issue.

Below, we will discuss three important responses to various external or internal aspects of the distribution factor: dependency theory; appropriate technology; and 'basic needs'. First, however, we need to look at some basic debate and evidence on the matter.

It is well known that income inequalities are pronounced in developing countries, and the consequences for those at the bottom end of the scale are severe indeed. Presumably for this reason, writers on the subject are prone to lapse into moralizing, or general statements of good intent. Whether one's intentions are good or not, there are some well-known arguments which come from orthodox, neo-classical economics which should not be ignored. Where a high rate of saving and investment is prescribed for development, a government may have to resort to incentives for people to save. But saving is a form of delayed consumption, and delayed consumption is hardly ever an option for the poorest groups in LDC's. Thus, the incentives, as rewards, accrue elsewhere. Even where exogenous assistance can obviate this requirement, it has been argued that aggregate increases in production will be most efficiently gained by directing technology, capital and other inputs at those believed to be most capable of using them, and this may not include the poorer groups. Even within the agriculture sector, which is usually where the great majority of poorest people will be located, such writers as Myint (op. cit.) occasionally come very close to saying that many developing countries cannot afford to adopt the more equitable forms of land use practice, until the basic matters of productivity and surplus have been attended to.

However, neo-classical orthodoxy can only be taken so far, before the <u>effective demand</u> of the population as a whole has become so undermined that significant growth is put under threat. This is no mere postulation: there is ample evidence, in many of the poorer LDC's, of growing impoverishment of a significant proportion of the total population. We have already seen, in our brief review of the Bangladesh economy in the previous chapter, that rural poverty is not static, but growing. It is worth examining this situation a little more closely, not merely as an empirical demonstration of the trend, but also to explore some of the reasons behind it.

The Government of Bangladesh, in a review prepared for the FAO World Conference an Agrarian Reform and Rural Development in 1979 (Ministry of Agriculture and Forests (1979): this article will henceforth be referred to as the WCARRD Review), is unequivocal in arguing that much of rural development is frustrated by highly inequitable distribution of land; the existence of widespread tenancy; competitive factionalison in the villages; and hierarchical patron - client relationships. Unlike many other rural societies, in Bangladesh villages the individual households are highly competitive, and superimposed on this are long-established power groupings which will effectively inhibit community oriented projects. Fragmentation of holdings through the Muslim based male inheritance laws (which compel division of a man's land amongst all his surviving sons upon his death regardless of his wishes), and the constant process of small landholders - under pressure of poor seasons and debt - selling out to larger farmers (who are often also the important money lending agents), are features of instability in the rural area.

One consequence of this situation is the burgeoning problem of landlessness. In the 1950s, about one third of the rural population of Bangladesh was landless. We can establish the more recent levels by reference to data on household size and landholding from the Statistical Yearbook (BBS (1980)). By converting the household landholding figures to a per capita basis, we obtain the following distribution from these figures:

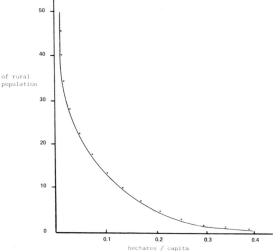

Figure 1. Per Capita Landholding Distribution in Bangladesh. Source: Derived from Table 4.3, 1979 Statistical Yearbook of Bangladesh, (BBS (1980))

The commonly accepted definition of 'landless' in Bangladesh is a household holding of less than about 0.2 ha. Using the above distribution and the demographic data, we can deduce that more than 50 per cent of the population is now landless, and over 70 per cent has a per capita holding less than the weighted average figure of about 0.1 ha/capita. Whether or not the rate of population growth slows significantly in coming years, there are still going to be many more people in the country than there are now and, on current indications, the proportion of them in the landless category will continue to rise.

The tenancy structure which results from this large scale landlessness tends, in Bangladesh, to be unstable. The WCARRD Review claims that some 70 per cent of sharecroppers cultivate the same piece of land for an average of less than three years. Sharecroppers and tenants presently have no rights to acquire land *via usufruct* in Bangladesh.

The distributional consequences of the land ownership pattern in the country are obvious enough. However, agricultural production overall in Bangladesh has, as we saw in Chapter 1, not kept pace with population growth over recent years. We have argued there that a principal reason for this probably lies in the pattern of development capital allocation amongst the major sectors of the economy but, at this point, it is worth looking at the possibility that a contributing factor is the pattern of land ownership and use.

Zaman (1973) has argued that for Bangladesh, the share-cropping tenure is efficient. However, Jabbar (1977) suggests the possibility that input and cost sharing tendencies measured prior to Independence by Zaman have changed, and he produces case study evidence that small farmers and owner - operators are the most efficient land users in Bangladesh. Hossain (1974) has arrived at the same conclusion, from different case study data.

On an intuitive basis, we might expect production to be lower with a highly inequitable land ownership and social structure. The WCARRD Review observes that, where programmes for co-operative development have been introduced in Bangladesh, they have frequently resulted in capital works of high private benefit to village elites, which are composed of large farmers, but low social benefit.

Indeed, the general effects of the so-called Green Revolution - the use of commercial fertilizers in addition to natural ones; adoption of improved high yielding crop varieties; irrigation and flood control technology; plant protection against disease; and so on - have been disappointing. Various explanations have been offered for this: Ahmad (1977) observes that technical deficiencies in the fertilizer mix resulted in poorer yields than expected from the tropical soils. A common explanation for failure of the 'technology option' for agriculture in countries with many very poor farmers is that in the absence of effective schemes to distribute the new inputs at low or zero cost, many potential users were simply unable to obtain them. Ahmad (1980) has argued that in Bangladesh, the introduction of this technology has systematically <u>worsened</u> the plight of small farmers, (presumably by lowering output prices, and making small operators more vulnerable to debt encumbrance). On large farms, operated by share-croppers, he suggests the effect of the new technology has been minimal, for at least one very good reason : share-croppers, under current arrangements, bear all the input costs in the crop growing, for which they receive half the output. But half the additional output from high yielding varieties is not sufficient to pay the additional cost of inputs needed, so the option is avoided and, from the share-croppers viewpoint, very sensibly so. Ahmed argues strongly for land reform, and criticizes the Government for not imposing it.

Land reform has in fact been attempted from time to time in Bangladesh. In 1950, legislation was introduced which limited landholdings to 33.3 acres per household. This limit was raised to 125 acres in 1962, in an attempt by the military government to create a progressive class of rural capitalists. In 1972, reversion to the 33.3 acre limit was enacted, and a deal more action on this matter was promised (or threatened, depending on the view of reform taken). Throughout the whole process, it seems (according to the WCARRD Review) that less than 2 per cent of land actually was caused to change hands, and virtually none of this went to families classified as landless.

It is probable that effective redistribution of land ownership in Bangladesh will not be possible, in the forseeable future : even if a revolution, genuinely prosecuted by the rural poor, did take place, things might not change as much as expected. In a sense, Bangladesh has already had two political revolutions - one at the time of Partition, the other in 1971, and the latter one did instal a government of an undoubtedly socialist persuasion. As we have seen, little changed for the rural poor. So far as we can see, with little or no concerted political activity of a pre-revolutionary nature (the creation of active cadres etc.) occurring, and the non-existence of a nascent revolutionary administration, there seems little prospect of a revolution genuinely capable of transforming the nature of society in the near future. But there is, obviously, the strong likelihood of continued political and economic instability in the next generation or so, if nothing is done to arrest the trends of landlessness and impoverishment. And if the political activism that results is directionless and unproductive of radical change, as we suggest will be the case for some considerable time, it will be no less disruptive of the development process for that.

Until and unless an effective revolutionary situation comes about, fairly gradual and subtle redistributive measures are probably the only reasonable course of action. The Government of Bangladesh is seeking to pursue such an approach through the Gram Sarkar (Village government) and Swanivar (Self-reliance) movements. Bangladesh has 68 000 villages, and the Government's contact with these is rudimentary. The village governments sponsored by the Government will, in the first instance, differ little in their effect from the existing, unofficial village elites. In time, as more people realise the process is an electoral one, and can be utilized to change things, more responsible and equitable administrations may result.

It seems fairly clear from the foregoing discussion that, so far as individual development projects are concerned, those that are oriented towards to purely production criteria, will rarely be value - neutral in terms of their distributional effects in Bangladesh, <u>even if</u> they have been conceived on a neutral basis. Briscoe (1979), in an analysis of energy use and social structure in a Bangladesh village, remarks that 'modernizing agriculture in Bangladesh is both essential and, <u>given the</u>

present social and economic structure, disastrous for the majority of the population.' Bangladesh cannot sacrifice growth - particularly when we examine present aggregate production and nutrition levels. Nor, however, can it purchase growth at the cost of not ameliorating the serious impoverishment of the landless. This affords us an important criterion for judging projects - in forestry or anywhere else: plans or programmes which do not recognise these requirements as an unsplittable pair are, almost by definition, themselves part of the problem. We have already seen some evidence of failure of the aggregate production approach to development and whilst there undoubtedly are some technical, infrastructural and administrative reasons for this, underlying them all, we suggest, is the failure to efficiently apply the one abundant source of capital - human capital - to the productive process.

The Bangladesh case represents perhaps the extreme situation, where such a significant proportion of the population has already become incapable of creating and sustaining effective demand for virtually any product, (even the most basic), so that the whole process of development is stalled. But other countries towards the lower end of the income scale must ask themselves whether they can allow trends in impoverishment and landlessness to continue (regardless of the overall aggregate growth figures being recorded). One wonders, for example, whether Mrs Ghandi's argument that, for the time being, nothing can be done for the bottom 20 per cent of India's population, might sooner or later need to be amended to 30 per cent, 40 per cent and ultimately a majority of the population - if something is not done fairly soon.

There is occasionally a tendency, when faced by problems of this magnitude, to cite the example of the Peoples' Republic of China, because that country has managed to combine acceptably high rates of growth with increases in social and economic equity (although Marrama et al. (1979) have shown that even in China, not all forms of income inequality have responded to the massive transfers that have been made. Urban bias in particular has remained stubbornly resistant to such measures). This book is not intended as a political tract, and the debate on the Chinese example will not be joined here, except for the suggestion that in this, as in so many other things, China may be unique. The high degree of social order and organization that followed revolution in China is not a general feature of LDC's. Nor, for those who tend to see solutions to LDC problems in simplistic political terms, is it an automatic consequence of revolution, no matter how violent or extended the revolutionary process may be. Also, China, for all its huge population, is also a very large land mass, and it is fair to say that in 1949 considerable potential for extension as well as intensification of agricultural production existed (and still does).

Systematic evidence on the effects of distribution in the developing world is not easy to collect. As can be implied from the discussion on population in Chapter 1, it would be most unwise to attempt to infer much about it from demographic trends and figures. Ahluwalia (1974) has attempted to measure the relationship between growth and inequality. First, he demonstrates that, not surprisingly, countries of socialist persuasion tend to have more equitable distribution: (about 25 per cent of GNP goes to the bottom 40 per cent of population in these countries, whereas in other developing countries this figures range from 9-18 per cent). On the basis of the limited data on income distribution over time available, Ahluwalia concludes that there is no strong pattern relating changes in income distribution to rate of growth of GNP. This is an interesting conclusion, because it suggests that there is little basis for the view that higher rates of growth must inevitably generate greater inequality. Ahluwalia suggests, however, that a generalized result of this sort is no useful guide for individual country cases. In searching for the relevant determinants in a given country, he advises that factors such as the concentration of wealth (including land) the mechanisms perpetuating it, and the existence of discriminatory market mechanisms should be included in the investigation.

Leipziger and Lewis (1980) have provided some statistical evidence from 38 countries in the low to middle income rankings of the LDC group. For the low income group, they suggest, the social indicators which they put forward as a quality of life index (life expectancy, infant mortality, literacy, per capita protein consumption, percentage of starch in diet, school enrolment and medical personnel per capita) are significantly and positively correlated with aggregate income growth. The authors' conclusion, for low income countries, is that the scope for improving welfare through redistribution in these countries is limited,

unless overall income growth is also occurring. They suggest that
concentration on achievement of patterns of growth which specifically
provide employment and income amongst the poorest group in the society is
the optimum policy.

So far as development analysts and aid agencies are concerned, the
relationship between development assistance and absolute poverty is of
major interest. Crosswell (1981) has studied this matter by examining
the distribution of poverty amongst LDC's, the relationship between
economic growth and poverty alleviation and the distribution of
development assistance amongst recipient countries.

In the process of this investigation, Croswell adduces sufficient
data and analyses to seriously question the postulated correlation
between economic growth and impoverishment - the so called 'Kuznets
analysis'. According to Croswell, findings such as this have tended to
confuse distribution between individuals, and distribution between
countries. On the surface, the rise in both LDC per capita GNP's, and
numbers of people living in poverty might seem to support unequivocally
the idea of inequality as a consequence of growth. But Crosswell
disaggregates the growth data on a country basis.

Table 2.3: ECONOMIC GROWTH AND POVERTY

Country group	Average per capita income	Incidence of poverty	Share of LDC poverty	Average per capita GDP growth p.a.	
				1960-70	1970-77
	$	%	%	%	%
Low income	170	50	81	1.5	0.9
Middle income	1 140	19	19	3.7	3.5

Sources: As in Crosswell, Table 2.

This table shows that in the LDC's where the bulk of the poor live -
the low income countries - economic growth has been extremely slow. In
the middle income countries, the incidence of poverty is in fact quite

low. Crosswell is also critical of cross sectional studies on income distribution, and, he cites some data which echoes our argument on population figures in Chapter 1 : that when time series data within a given country are used instead of cross sectional data, the implied relationships are different; in this case, they show that there is no trend between changes in inequality on the one hand, and any initial inequality; level of GNP; or rate of change of GNP on the other.

Does this mean that the current emphasis in the literature on 'basic needs' (see following Section) policies as a means of averting intensifying poverty is misplaced? In Crosswell's own view, the answer is in the negative : the development community is still faced with the problem of alleviation of poverty where it occurs. The interesting addition which comes from Crosswell's findings is the inter-country distribution. He shows that, whilst most poverty is concentrated in low income countries, the bulk of concessional foreign assistance has been directed elsewhere: to the middle income countries. He provides the following interesting data in support of this contention.

Table 2.4: DISTRIBUTION OF POVERTY AND OFFICIAL DEVELOPMENT ASSISTANCE(a), 1976-1978

Recipient country per caput GNP (1977)	% of world's total poor	ODA commitments (10^6 p.a.)	% of total ODA
$150	17.4	3 385	13.3
India ($150)	37.2	2 100	8.3
$150-300	26.1	5 793	22.8
$300	19.2	14 125	55.6
Total	99.2	25 403	100

(a) As defined by OECD: grants or loans by the public sector (including international agencies) that have as their main purpose the promotion of economic development and welfare, and are offered at concessional terms.

Sources: As in Crosswell, Table 3.

Crosswell's data and analyses strongly support the idea that economic growth is not synonymous with intensifying inequality. The remaining question, then, is whether development assistance actually aids the growth process. Crosswell's review of results on this matter shows them to be not conclusive. A general result which suggests a negative conclusion is typified by Weisskopf, who was able to show, over a certain range, that foreign capital inflows and domestic savings rates are inversely correlated. But there are arguments about the interpretation of this result: it may not necessarily mean that foreign inflows (including assistance) are substituting for local savings. It may be that the some exogenous factors which cause local savings to fall - war, weather etc. - are those which stimulate rises in foreign aid. Crosswell also cites some data on growth and aid and other fund inflows in a group of Asia's countries which show strong correlations between aid inflows and growth. At the present time, the overall impression is one of ambiguity - it is not possible, on the basis of existing data and methodology, to determine the causality between growth and aid.

2.3.1 Dependency theory

An important explanation which has been offered as to why LDC's have not progressed - either in terms of their trade with Centre countries, or in their internal developments - is a body of literature which we can generally term Dependency theory. It is typified in Frank (1978), Cardoso (1979) and Amin (1976). In essence, it grew out of general neo-Marxist concepts that have been powerful influences in development economics since World War II.

The essence of dependency theory is, very broadly speaking, that the penetration of capitalism into under-developed countries is not a solution to, but rather is the cause of, their problems. Amin, for example, argues that the central determining contrast between developed and under-developed countries is to be found in their production structures; the former are 'self-centred', based on the development of a domestic mass market for consumer goods. Under-developed countries by contrast, are externally oriented, depending heavily on exports to fund the local market for 'luxury' goods demanded by various 'parasitic' social groups in the country: rich peasants, merchants, state bureaucracies and so on. These groups, according to the dependency theorists, encourage the entry of imperialist exploiters from abroad, with whom they then proceed to share the spoils, whilst the country's development suffers. Capitalism thus develops as a result of external forces, and it will serve the interests of those forces: development of the export sector occurs because its production costs can be held lower than those in developed countries - but the profits made will be repatriated to the developed areas. Wages will be held down, through the alliance of foreign investors and local vested interests.

Dependency theory became prevalent largely as a result of observed failures in the 'modernisation' approach to development. As we have already seen in 2.1 above, it became fairly obvious soon after the War that many under-developed countries were not being integrated into the global market - specializing (first in primary production, but soon in industrial products) and so on - as the original theories on development argued they would. This failure led, very often, to import-substitution policies in many places, as countries attempted to intervene in the trade cycle, specifically to encourage the transformation of factor endowments into an industrial base. All too often, these policies resulted in overall reductions in economic welfare, and a particularly severe diminution in the living standards of lower order workers and peasants.

Two schools of thought developed at about this time. Firstly, the 'free trade' school, which argued that the experience showed the folly of state intervention - except, as Johnson and others of that ilk would argue, where intervention is necessary to ensure the economy responds correctly to international price signals.

Dependency theorists, on the other hand, argued that state intervention in trade certainly was necessary, but that it was not enough. Obviously enough (from a reading of some dependency literature) this sort of argument has led to many extremist interpretations. But the central argument is one which warrants some thought: the integration of weaker economies into a system dominated by powerful, technologically advanced ones leads to an asymmetrical system, where the weaker economies become dependent on the stronger, rather than genuinely competitive with them. One version of dependency theory contrasts 'competitive capitalism' (the model postulated for developed countries) to monopoly capitalism (which comes about in under-developed countries, as foreign sectors are 'grafted on' to local economies). Under competitive capitalism, technical change brings about falls in product prices, therefore rises in demand, therefore rises in demand for labour - precisely the result, we would agree, that most under-developed countries would want. But monopoly capitalism produces a different result (this reasoning, incidentally, seems based on microeconomic behaviour as presented in standard theory of the firm analysis for enterprises within a given economy): technical change will not produce price falls, but rather leads to increased profits and higher wages within the industry concerned. This leads to a privileged labouring class (therefore, a vested interest) within the export sector, and a stagnation in overall demand for labour. This is, as one can see from general observation of many under-developed countries, precisely the situation that many of these countries are in today.

But is the explanation offered by dependency theorists correct? If so, how do we explain the fact that many previously under-developed countries <u>have</u> progressed, in the modern context - the newly industrialised countries? This, as so often seems to be the case in the development economics area, leads to an enormously complicated and voluminous debate in the literature. The best we can do here is refer very briefly to two modern articles on the subject, which represent ends of the spectrum of interpretation of events of the 1970s.

Bienefeld (1980) argues that, whilst transformation of the wage structure for most of the population has occurred in some countries (e.g. Singapore), there are many more countries where this has not happened. However Bienefeld does point out that rapid overall growth has occurred in many countries, as a result of dependent development: he points out that it is no longer possible to be as definite as were some of the early dependency theorists on the limits of capitalist development in dependent countries.

Schiffer (1981), in a trenchant criticism of the economic ideas of Amin, argues that the international spread of capitalism has not been limited to luxury goods in developing countries, and that industrial growth in these countries <u>has</u> been based on broad expansion of domestic markets.

Schiffer admits the probability that industrial development in under-developed countries has been dependent on multi-national corporations in many places, and that deep divisions in society have been created by the process of change. But he specifically rejects the idea of a permanent group of poor, 'marginalized' markets being the result. He argues that this situation is temporary, and points to evidence which suggests that in such countries, more and more people are being integrated into capitalist structures. He suggests that writers such as Amin tend to blame 20th Century economic imperialism for food and population problems in under-developed countries - he blames the combination of modern health and sanitation programmes (reducing death rates), and persistence of pre-imperialist land tenure systems in the agriculture sector. His basic philosophy seems to be that the large discrepancies that arise between productive activities are normal for capitalist development.

This argument, in effect, returns us to our initial observations at the beginning of Section 2.3 above; one must make a judgement about the extent to which very low income countries in particular can allow effective demand to be reduced, in pursuit of uneven growth policies. In our view, the more purist form of dependency argument, which has it that countries which allow capitalist penetration and an external orientation in their production behaviour <u>cannot</u> progress, is in retreat: the empirical evidence suggests otherwise, and indeed some of the most successful economic transformations (accompanied by significant reductions in <u>general</u> poverty levels) have been based precisely on this <u>modus operandi</u>. But, on the other hand, for very poor countries, the socially disruptive effects of dualistic economic policies are significant, and in such cases many of the observations made by dependency theorists are relevant.

2.3.2 <u>Appropriate technology</u>

Modern technology in developed countries has evolved under conditions of labour scarcity, and a relative abundance of capital. In most LDC's, the situation is reversed - labour is abundant, and capital scarce. Technology, and methods of production which have been designed on the basis of capital intensity will, in the LDC environment, be the least efficient option. There are, moreover, a large number of practical reasons why introduction of capital-intensive methods, which tend to be complex, into LDC's will be difficult: some of these were discussed in Chapter 1 in relation to Hagen's observations on this subject. The requirements such methods impose for highly specialized skills, a sophisticated supply infrastructure, and a comprehensive repair and maintenance network all create severe problems in LDC's.

It may be wondered why this subject has been included under the general income distribution head in this book. The principal reason for this is because, in our view (and it is one which has been argued by Singer (1975), among many others) there can be a strong connection between technology and income distribution. The connection comes about because of the _nature_ of the technology which has been transferred. LDC's are required to create employment at much faster rates than developed countries, to absorb their rapidly growing populations until such time as the rate of growth of population can be slowed. But, by definition, the resources they have (on a per capita basis) to achieve this task are only a fraction of those available to developed countries. Utilizing the _same_ technology as developed countries (which, these days, themselves have problems in creating sufficient employment to absorb additions to the labour force) will therefore not allow LDC's to create anything like the number of jobs needed.

But LDC's face a dilemma: to break away from inappropriate, Western technology requires the invention of technology which _is_ appropriate to the given country's needs. Typically, LDC's spend something in the order of 0.1 per cent of GNP on research and development, compared to about 3 per cent in developed countries. To put this another way, something over 90 per cent of all research and development expenditure in the world is incurred in developed countries, and is oriented almost totally towards their own production situations. The U.N. estimates that, at best, LDC's could spend 0.5 per cent of GNP on research and technological development; institutional and training constraints would, initially at least, preclude more, even if the funding were available.

According to Singer (op. cit.), the 'brain-drain' is the clearest sign of the effect that concentration of science and technology in developed countries has on LDC's. Scientists and researchers leave poorer countries for the developed area, attracted by facilities, salaries, and, not least, the opportunity to do the sort of fundamental and esoteric research of which Nobel prizes and other accolades are made.

Further, Singer argues, the <u>internal</u> brain drain may be even more significant: that is, the tendency for LDC scientists to <u>behave</u> as members of the international scientific community, addressing themselves to problems it deems as important, even when they do not physically leave their country.

The remedies to the technology problem - if it is recognised as a problem (and it must be noted here that very often it is <u>not</u>, within LDC's, notwithstanding what may be said about it in the academic literature) - are not simple. Singer argues that an expansion of research and development expenditure in LDC's - even if only to the 0.5 per cent limit - might carry the creation of science and technology beyond a threshold below which ineffectiveness and the brain drain predominate. But, since no-one seems to know precisely where this threshold may be, we are left to wonder whether Singer's contention is anything more than pious hope.

A much more fruitful possibility, it would seem to us, would be to alter the nature of science and technology in LDC's, away from attempts to duplicate fundamental research institutions, and towards a capacity to <u>identify</u> research or technology needs, and <u>evaluate</u> the responses made (under contract in the developed world, if no other avenue is available) elsewhere. There is no reason why developed countries - or international aid agencies - should not at least investigate the prospects of diverting some funding away from the notoriously difficult task of more research or technical institution building in LDC's, towards use of more of the massive research capacity already installed in the developed world for solution of specific LDC problems. To some, this may not be a popular idea, and may smack of self-interest on the part of aid donors - but the only relevant test of it will be whether or not it can deliver well-designed, low cost technology for priority purposes in LDC's better than any other feasible system. Certainly, in our view, it has more potential than either the continued indiscriminate transfer of Western based technology, <u>or</u> the attempt to create <u>complete</u> research and development infrastructures within LDC's. It is important to bear in mind in this context the view of writers such as Vaitsos (1970), who argue that the type of technology needed by developing countries very often <u>is</u> amply available; it is the means of deciding upon it, bargaining for it and acquiring it, which are presently the problem areas.

It is not our intention here to completely exclude all other options in the appropriate technology area: rather, we are concerned to give some weight to the direct approach of importing appropriate technology - an approach which seems to be frequently ignored. There are, of course, many courses of action available to a government which wishes to improve the performance of technology in the economy. Marsden (1970), for example, lists no fewer than 24 steps which might be taken, ranging from fiscal measures to encourage local innovation, or discourage unnecessary importation, through to creation of extension, library or education facilities in the technological field.

As a final observation on this subject; it is essential to bear in mind that inappropriate <u>means</u> of production are only part of the problem. In the common LDC situation, where urban elites (of which the scientific community will be a part) are insulated by status and wage levels from the remainder of the population, it is likely that the wrong sorts of product will be created. In this sense, technology becomes the vehicle by which dualistic economic trends proceed. We will return to this subject again in the specific forestry context in future chapters: it is one which, in the often frantic search for better ways of doing things, is often disregarded. Thus, when considering the employment of technology (whether domestic or imported) in industry, LDC decision-makers would do well to consider the role not only of the process, but also the product:

> 'industry should produce the simple producer and consumer goods required by the people, the majority of whom live in the countryside: hoes and simple power tillers and bicycles, not air-conditioners and expensive cars and equipment for luxury flats.'
>
> Streeten, 1977

2.3.3 Basic needs

We have left the basic needs issue to last in this Chapter because the strategy involved in it, and the discussion of it in the literature, is in fact the current framework upon which much of the distribution debate rests, and much of what we have discussed already in this Chapter is relevant.

The basic needs approach has grown out of the experiences with growth and income distribution in development programmes. As Hicks and Streeten (1979) have pointed out, it was a response to the observation that in some LDC's, highly concentrated and unequal growth prevailed for long periods with little indication of a spreading of the growth effect. Since no convincing evidence could be found to suggest that inequality was needed for growth (and indeed, as we have suggested in this book, might eventually approach the stage of being an inhibitor of growth), the idea that something specific should and could be done for the bottom group in LDC's grew. What the basic needs strategy involves, as its name implies, is a 'safety-net' approach: a level of welfare below which individuals should not be permitted to go. The essential basic needs are considered to cover six areas: nutrition, basic education, health, sanitation, water supply and housing. A great deal can be (and has been) said about how these parameters should be measured and defined, and what the limiting levels for them should be. However, our interest here is less with these technical issues, than with the concept of basic needs as an integral part of development strategy.

The basic criticism of the basic needs approach encompasses arguments which we have already discussed in this chapter, to the effect that there is an inverse correlation between growth and redistribution, and that LDC's must choose growth. As we have seen, there is still considerable controversy about this matter. Hicks (1979) has attempted to answer the question of whether there is a trade-off to be made between growth and basic needs. Using an econometric approach (which, as Hicks himself suggests, leaves the matter of causality uncertain), the findings from this analysis are that: countries which have made substantial progress in meeting basic needs over the 1960-1973 period have not recorded lower GNP growth rates; and that the attainment of a higher level of basic needs satisfaction appears to lead to higher growth rates in the future. In Hick's view, although the conclusions of the study are of a preliminary nature, it seems likely that countries generally do have the capacity to meet basic needs without crippling other programmes aimed at growth.

Hicks does observe, however, that his analysis does _not_ explain _how_ countries meet basic needs without reducing growth, and _why_ some do rather better than others in meeting basic needs.

Dell (1979) argues against the basic needs strategy, on the grounds that it is not a development strategy at all. Although Dell bases much of his criticism on the familiar argument that LDC's must give high priority to the transformation of their economies along industrial and technical lines, he also brings out some practical difficulties with the basic needs approach. Referring to an unnamed Asian country, he points out the consequences there of a project to upgrade provision of water, sewage facilities and other improvements which were aimed at poorer groups. The immediate effect, he suggests, was that the value of land and housing in the serviced area rose, the impoverished occupants felt compelled to sell out to richer people and move on to slum areas on the outskirts of the country's capital city.

Along similar lines, we have already remarked in our discussion of the Bangladesh case at the beginning of this chapter, on the common result of supposedly co-operative development schemes in fact delivering maximum benefits to richer groups, and little or none to poorer people. One fairly well known example involved the provision of funding and facilities for the digging of tubewells and establishment of pumps for irrigation. The intention was that the wells would serve a command area which would at least involve small, poor farmers in distribution, and location of the systems was planned accordingly. However, it eventuated that the great majority of the wells ended up on the lands of wealthy farmers who either over-used the water themselves, or sold it to smaller farmers - resulting in either case in negative effects on both equity and efficiency.

Such examples are not really arguments against the concept of basic needs, but rather are indicators of the difficulties of transferring income opportunities directly to poorer groups in rigid, hierarchical rural societies. Dell argues that international aid agencies, by not understanding the real nature of such problems, can make things worse by attempting to intervene on the distributional side. They should concentrate, he suggests, on the technical input areas where their expertise lies. We would certainly agree that a great deal more sociological and cultural research should precede rural development projects which have some redistributional elements in them. But, as we have already suggested in this book, there is really very little spare ground between the sort of project which discriminates in favour of the poor (at least to the extent of ensuring they retain _some_ proportion of the benefits), and those which actively widen the gap between them and the rest of society. Projects which seek to be value-neutral, or 'aggregate' in the nature of their benefits will, as we have seen, usually end up directing benefits away from impoverished groups. As we have already noted, there would not seem to be any reason to conclude that inequality is an inevitable artefact of growth. Given that the logical consequence of large scale impoverishment (which is the reality for many low-income LDC's, and, incidentally, was _not_ the case for many middle income countries which are presently developing rapidly) is a collapse in effective demand, our conclusion is that the orientation of at least a significant number of development projects towards the concept of meeting a minimum set of needs among the poorest groups, is justified. In fact, it may be that the only real requirement under the basic needs strategy is to provide significant income opportunities to poorer groups: the associated needs of nutrition could then be taken care of by the individuals themselves. Health, sanitation and education programmes, it seems to us, can be justified on aggregate grounds as being highly productive of economic progress – such activities do not _require_ a basic needs justification.

NOTES (Part 1)

(1) 'In fact, it was precisely the inequality of the distribution of wealth which made possible these vast accumulations of fixed wealth and of capital improvements which distinguished that age from all others..... If the rich had spent their new wealth on their enjoyments the world would long ago have found such a regime intolerable. But like bees they saved and accumulated not less to the advantage of the whole community because they themselves held narrower ends in prospects'. (The Economic Consequences of Peace.)

(2) See, for example, the reference to Ahmed (1980) in Section 2.3 of Chapter 2.

(3) President's address to the Board of Governors of the International Bank for Resources Development, October, 1976.

(4) Averaging their 'loss' figures gives us an annual figure of $5 billion - well under $2 for every person presently living in LDC's.

PART 2

THE ROLE OF FORESTRY IN DEVELOPMENT

".... their forest and forest industries will be subordinated to, and carefully geared to, their national development priorities. And among these the most imperative is to ensure their people are adequately fed. This is not simply a matter of switching investment from industry to agriculture. It is a matter of facilitating and encouraging those structural changes which will enable the rural masses, at all levels, to feed themselves, and to move progressively beyond that to the production of an agricultural surplus which will ensure that the urban population, too, is fed, and no longer dependent on food air or on costly food imports. This is the only basis for sound industrialisation. Such governments will therefore closely scrutinize all present and possible future forestry activities from the standpoint of how best they can protect, support, promote and diversify the agricultural economy."

J.C. Westoby
(8th World Forestry Congress)

Chapter 3

APPROACHES TO FORESTRY DEVELOPMENT

Forestry can be practiced in a variety of ways. At one extreme, there is the large scale plantation and natural forest management by a traditional government department, for purposes (principally) of supplying material to a highly capitalized, technologically oriented processing sector, or to export. At the other, there is small scale rural forestry, practiced at the household of communal level by rural dwellers for purposes of fuelwood, basic structural materials, shelter, food and fodder and the range of minor forest products. Superimposed over the former type, in LDC's, is very often the slash and burn practices of shifting cultivators who may or may not have been the historical residents of the forest areas. This phenomenon, and the associated one of unofficial encroachment of forested lands for permanent agriculture, are the most visible manifestations of the land use (and, ultimately, the economic) conflicts and paradoxes that so often characterise the forestry sector in LDC's.

As we will see in the following brief review, there is as yet no clear concensus as to how the newer ideas on development should be put into forestry practice in LDC's. There is, however, more agreement that what was written about the role of forestry in LDC's in the 1950's and 1960's - and even well into the 1970's in some quarters - was (and is) very largely inappropriate. Nevertheless, large elements of the older view still persist in the forest sector administrations of LDC's. We need, therefore, to examine some of this older material, firstly to understand why it does not stand the test of critical scrutiny on the basis of more recent general development precepts, secondly as a basis from which to examine why much of it survives in practice, and thirdly so that we can consider some measures for rationalization and adjustment that can be taken in practice.

In addition to the literature on the role of forestry in development per se (which in fact is not extensive) there is a deal of material on the specific techniques of economic evaluation of forestry projects in LDC's. The general literature on project evaluation in LDC's, upon which the specialized forestry material is based, is very large. Although in theory this material is not of direct relevance to our subject, we do need to deal briefly with one or two aspects of it, for reasons which will become apparent.

3.1 Project Evaluation in Forestry

It is, or should be, a commonplace amongst practitioners and theoreticians alike in the project evaluation field that the results and conclusions from its application are highly sensitive to assumptions made about valuation of the various costs and benefits and externalities involved. These, as will probably be already apparent, are in their turn dependent on the basic concept or model of the developing economy used. Whilst project evaluation is not intended as a means of deciding upon the broad issues in development, it is often used for this purpose, either because of lack of definite guidelines on the general development strategy to be employed in the economy in question, or because of a lack of awareness or understanding of such guidelines on the part of the analysts and administrators involved. Arnold (undated), in identifying this tendency in the analysis of forestry projects, observes that, in such cases, "the project's relationships to such other objectives as employment, balance of payments, distribution of income, etc., has been confined to listing or stating the number of jobs created, foreign exchange earned or saved, etc., by the project or project design chosen on the basis of its expected economic performance. Thus the project which emerges is not necessarily the one which represents the best balance in meeting the various objectives." Arnold notes that the sequence of project preparation from its origin of the technical levels through the administrative and financial stages to its economic appraisal, is very prone to the problem of the analysis being narrowed or constrained to the extent where the best overall alternative may be prematurely discarded, or never considered. We would extend this argument to the overall sectoral level: if sectoral activity is composed of a series of projects or programmes which are evaluated through the same sequential system, then the sector strategy as a whole will be unnecessarily constrained, and prone to errors of omission, unless specific steps are taken - usually from above the sector administration levels - to alter this approach. The most common omission, we suggest, is adequate consideration of income distribution : sectoral project evaluations often either ignore it, or attach neutral weights to consumption gains, no matter where these accrue.

We should by now be in no doubt that distribution matters. Even when analysing the <u>difficulties</u> in the theory of welfare economics, Little (1973) is quite clear on the point that a judgement about whether a particular distribution of income is good, or bad, <u>must</u> be made by someone. Little and Mirlees (1974) are highly critical of practitioners of cost - benefit methodology who overlook this requirement. Of the UNIDO (1972) guidelines for project analysis in developing countries (which advocate the use of aggregate consumption as the numeraire in calculation of accounting prices) they say "it is not the case that aggregate consumption is the ultimate objective, which might make it a good numeraire. On the contrary, the authors emphasize that the consumption of different economic groups should be given different economic weights." The reasons why this is so go well beyond the moral imperative, as will be seen from our review in Chapter 1: where impoverishment and destitution are sector and group specific to an appreciable extent, and where the worst affected groups are potentially important in terms of productivity, then re-distribution is a basic ingredient of, rather than an adjunct or side benefit to, aggregate income growth on a sustained basis. Yet, unweighted income growth criteria have been frequently used in the project evaluation literature. In Gane's (1969) detailed exposition on forestry and development in Trinidad, it is argued that forestry projects can be analysed on the basis of aggregate gains to consumption. Gane opens his arguments for this viewpoint by reference to classic analyses by Pigou and Little (op cit) which argue that comprehensive direct measurement of welfare cannot be made, and that welfare effects cannot be estimated without value judgements about income distribution. Gane seems to deduce from this that it would be preferable therefore to avoid assigning weights to consumption by the various economic groups. However, as we have seen, it is impossible to avoid doing so, and the weightings implied in the aggregate criterion are no less definite, (but are probably decidedly less desirable), than those that would result from a more subjective weighting. Nowhere, it seems to us, does Little argue that deliberate allocation of different weights for income changes for different groups in society is undesirable: indeed, he suggests it is unavoidable.

The interesting theoretical development which seems to have arisen from the work of Lipton, Streeten and others referred to in the previous chapter is that _positive_ discrimination in investment capital allocation towards the industry containing the bulk of low income people might, ironically, have led many LDC's to higher _aggregate_ income growth in any event. As we have already seen, whether such a finding would be repeated if applied to the lowest income groups _within_ that sector is as yet imperfectly researched, but the nexus between sustained development and maintenance of demand may eventually lead LDC's in that direction anyway. What Lipton and others seem, indirectly, to have demonstrated, is that the use of the aggregate consumption criterion would have been acceptable, had all the important opportunity costs involved been reasonably calculated or estimated. Gane returns to the matter of opportunity costs, in a more limited way, later in his analysis, but we can see now that his exposition would have benefitted by their inclusion at this point.

Gane develops his defence of the aggregate income criterion by claiming that forestry is unlikely to exercise _negative_ effects on income distribution. He suggests that the sector by its nature favours lower paid workers, and argues that the flow of forest products will not cause income reductions to low income groups, and through import replacement or export generation will create employment for them. We have encountered all these arguments in a general sense in the previous chapter. They do not in fact stand, when the bases of comparison are widened beyond other LDC industrial sectors: failure to remove this sectoral constraint is the basic conceptual flaw in material of this type. If the forestry sector, including the forest processing industries in an LDC, is consuming large amounts of development capital, then it seems, on the basis of Hagen's and Lipton observations, that negative effects on income, and income distribution, _will_ in fact be occurring, but this only becomes obvious when a wide enough range of possible alternative uses for such capital is considered. It is in fact extremely _likely_ that the flow of forest products, or at least of those manufactured by and for a very narrow socio-economic stratum of society, will reduce the income of poorer groups in society.

The application of project evaluation methodology can, as we have said, only be as good as the broad economic assumptions upon which it is based: it cannot substitute for them. Even in cases where general guidelines on the need to create employment, or re-distribute income are passed down from above the sector administration, the effect of these may be marginal, relative to the scale of change needed. It may be relatively easy for a sector administrator to adjust the existing programme of his sector, in accordance with the letter of such edicts. It will usually be a good deal less simple for him to perceive that the problem implied in such edicts might be far more profound - that a strong linkage between the existing economic <u>structure</u> of his sector, and the continual impoverishment of certain groups in society, might exist. In this, he will ultimately be guided by his received opinion of the appropriate role for his sector in the economy. This, if we accept the thrust of Lord Keynes' famous dictum(1) will be a version of what academics and theoreticians have said about it. Therefore, we will now turn our attention to what has been, and is being argued as the most appropriate role for forestry in the developing economy.

3.2 <u>Forestry in Development</u> : <u>The Original Westoby View</u>

The view of forestry in LDC's taken by forest economists and planners has, not surprisingly, been influenced by general concepts and thought on the economics of developing countries. Basically therefore, there have been two distinct phases since the 1950's - the earlier view, based more or less on the industrialization approach to development of the forestry sector, and the later approach, based on concepts of rural community forestry, village level industry and so on. It will be out objective here to examine critically how the general economic arguments have been translated in to forestry terms, with a view to arriving at a generally acceptable approach to forestry in LDC's.

As remarked in the Introduction to this book, there has been relatively little written on the specific subject of the role of forestry in poor countries. It is fair to say that the case for the industrialization approach to forestry in LDC's was made in its most complete form by Westoby (1962), in a long article. Until the general concept of development implicit in that work become subject to strong doubt and criticism, (and indeed, for some considerable time <u>after</u> this) little development of the arguments and contentions in it occurred.

Westoby delineates his area of interest by suggesting that forests provide a raw material for industries which have "acquired great importance in developing countries". He reinforces this constraint on his analysis by excluding fuelwood from consideration altogether : suggesting that it is of secondary importance from the point of view of economic growth. He does point out that the effect of fuelwood may not always be negligible, and suggests that South Asia is one area where it might assist agricultural productivity - a welcome exception indeed, (although not developed in his analysis), considering the fact that in most of the developing world fuelwood is the most important forest output for most people, and is destined to remain so for some time to come.

A central theme of Westoby's argument is revealed in the identification of the high degree of structural interdependence between forestry and the industrial sector of the economy. These strong forward and backward linkages with the remainder of the economy demonstrate, according to this argument, the seminal importance of forestry in the development process. Nautiyal (1967) has shown that in fact the forestry and timber based industries have quite <u>weak</u> backward linkages. In our view, from a reading of the original data of Chenery and Clark which Westoby used, it is not apparent that the forestry sector does have an impressively high degree of interindustry linkage with the rest of the economy at all. Even if it were so, the high degree of variability of multiplier results (which often seem as sensitive to the definition of the interindustry model used as to genuine sector differences), from one country to another (and even from one region to another within countries) make the applicability of it to the general development case highly suspect. Westoby uses a 'special case' argument - that of exclusion of the construction sector from intermediate demand (and therefore from 'multiplier effect') - to defend his argument that forestry multipliers <u>would</u> be high if this effect were included. Perhaps so, but similar arguments undoubtedly apply to other sectors, and until someone examines them all on an exhaustive and systematic basis, results for individual sectors are of very limited value.

In any event, as we have already seen from Hagen's analysis, in Chapter 1, a high degree of structural interdependence with industrial sectors of the economy might be the very last thing we want of a primary sector, in an LDC where constraints of technology, capitalization and infrastructure are critical. The creation of <u>demand</u> for intermediate inputs from a range of industrial sectors is no guarantee whatsoever that such demand will be met - not, at least, in the real world of the developing economies. In other words, the existence of strong forward linkages for a given industry, measured on an historical basis (usually from a <u>developed</u> country economy) tells us nothing about whether similar effects can and will arise in the particular LDC we are studying.

Westoby is on firmer ground when he suggests that income elasticities of demand for forest products are high, in low income countries, and that therefore any country with fairly rapidly growing income is going to demonstrate rapid demand growth for these products. However, apart from the possibily undesirable distributional aspects which apply to this (see the discussion of Gane's work, and the review of the Bangladesh situation in Chapters 1 and 2 above) it seems to be, once again, an incomplete argument : high income elasticities of demand undoubtedly characterize the incipient markets for a great many consumer items in LDC's - it would be most surprising to find this were not so. There is no systematic evidence that forestry products are preferable in this respect. Nor is there any compelling reason for us to accept Westoby's associated argument that this market can be catered for relatively easily in LDC's because the forest industries are either easy to set up, or generate such high rates of return that investment will be quickly recovered from them. Once again Hagen's argument that the early phase of industrialization <u>must only</u> attempt to deal with already existing industries in LDC's is relevant here. Finally, in considering this argument, we should perhaps remind ourselves that the principal task in the early stages of development is not to arrange the structure of growing demand, but to <u>initiate</u> income growth and its appropriate distribution.

The essential difference we have with Westoby is in the view taken of the dynamics of the development process. Westoby wants the forestry sector to lead the LDC economy where he believes it should go, and he is in some haste to get there. But the mainstream of present thought on development would suggest that evolution towards development will be slow, and must be phased from a basis of effective growth in rural incomes and output, rather than from a superimposition of a modern, industrial sector. In the early stages of development, it is not a gross over-simplification to say that economic activity which is not obviously part of the solution (which involves rural productivity and the reduction of absolute poverty) will be part of the problem. Despite the time lags inherent in forest production it is no more logical, in our view, to pursue accelerated establishment of the modern sector, or aggregate production as an end in itself in this sector, than it is to do so in agriculture, in the midst of the realities of intensifying rural poverty, a backward social structure and an ineffective capital market. The essential question is not when the forest resource should be mobilized (we would agree with Westoby that this must be as soon as possible), but how.

As noted earlier, little development of the ideas in Westoby's article occurred until the large shift in the general approach to development penetrated forestry circles. Thus, in FAO (1974) for example, we see reference to the favourable status of forestry as an industrialization base: the high growth of demand in relation to income growth; the strong forward and backward linkages of the sector with the economy; the ease of establishment of forest industries; the role of the sector as mucleus for industrial development; positive foreign exchange and employment effects. Much of the general approach in Westoby's original article also survives in MacGregor (1976), Von Maydell (1976), Gregory (1972).

3.3 The Forest Industries of Bangladesh

The forestry sector of Bangladesh is a reflection of the dualistic nature of the economy in general which arises from the 'industrialization first' approach: on the one hand there is the Government forest resource, for the most part occurring in large concentrations and managed principally to supply output to the large scale forest products industries. On the other, there is the village forest resource, which supplies the basic fuel and structural needs of the rural population, and operates more or less independently of Government activities. The situation that pertains in these resources will be of interest in our general discussions of the social, environmental and fuelwood aspects of forestry, and we will return to it in the next Chapter. Here, we are concerned with the evaluation of the forest industries of Bangladesh, in particular the larger scaled (principally Government owned) industries. This review draws heavily upon a report on the economic performance of this sector (Douglas et al (1981)); some of the detail necessary to establish our case is extracted from that report into Appendix B of this book.

Bangladesh is currently involved in pulp and paper production (from wood and non-wood furnishes) including rayon and cellophane; sawmilling; ply, veneer and reconstituted boards; and a range of secondary processes - furniture manufacturers, packaging firms, and so on.

Before attempting to draw some general conclusions about the forest industries sector, some (very broad) assessment of the major products performance is warranted.

3.3.1 Pulp and Paper

Paper making in Bangladesh is completely dominated by the Bangladesh Chemical Industries Corporation, a large Government conglomerate which controls some 30 enterprises, including a newsprint mill, two other paper mills, a pulp mill, and a rayon and cellophane plant. The basic economic performance data for these operations is given in Table 3.1 below, and in more disaggregated form in Appendix B.

Table 3.1

Economic and Output Data: Bangladesh Pulp and Paper Sector

	1975/6	1977/8	1979/80
Production ('000 mt)	39.8	72.5	79.1
Domestic Sales ('000 mt)	30.2	50.8	52.3
Operating surplus (Tk 10^6)	- 157.2	- 111.8	- 121.8

Source: Table B.1; Appendix B.

In each case, very rapid rises in imported input costs have occurred in recent years. In the case of the newsprint mill, foreign inputs now account for some 65% of total costs. Most of the plant established in Bangladesh is old, and is sub-economic in terms of scale and technology. Yet production from the mills far exceeds domestic demand, and much of the product from the mills is sold to export at prices well below domestic levels (and below production costs, in most cases). If foreign exchange earning (or saving) is seen as the principal benefit of the paper industries (as is frequently implied), then they are a remarkably expensive means of achieving this: in Douglas et al (1981), shadow price calculations are done which show that the cost of earning or saving one Taka of foreign exchange in the paper industries costs Bangladesh almost three Taka - this amount is well out of line even if significant over-valuation of the local currency is assumed.

Perhaps more than any other group of industries, the pulp based sector of Bangladesh illustrates the difficulties of attempting to establish capital and technology based industries in an economy of that nature, regardless of what benefits for domestic production may have been adduced at the time. As it stands, the paper-making industries are now high on the list of the largest money losers in the economy. The infrastructural and capital requirements to sustain profitability in industries of this nature are not available in Bangladesh, and will not be for the forseeable future. Raw material resource constraints are also apparent - particularly in view of the priorities for use of the country's dwindling resources as outlined in the next Chapter.

3.3.2 Manufactured Board

The fortunes of the manufactured board sector in Bangladesh are less uniformly bad than in the pulp and paper sector. As will be seen from Table 3.2 below, it is a much smaller sector overall:

Table 3.2

Economic and Output Data : Bangladesh Manufactured Board Industries

	1975/6	1977/8	1979/80
Production baseboard ('000 m^2)	1264	1858	1598
particle board (mt)	1684	3098	2476
ply ('000 m^2)	697	1500	1200
Domestic Sales h'board ('000 m^2)	1189	1756	1449
part. board (mt)	893	471	n.a.
ply ('000 m^2)	697	1500	1200
Operating Surplus (Tk 10^6) hardboard	-0.7	-0.1	+0.5
part board	-5.0	-6.8	n.a.
ply	-1.3	+0.1	+1.5

Source: Table B.2; Appendix B.

Both the hardboard and ply industries sell principally on the domestic market (the latter almost entirely to production of tea-chests for the export tea industry). Although both industries have a range of technical and other problems, they are at least managing to make ends meet, and to sell their products at a reasonable price.

The particleboard industry is another matter : presently, it is based on jute fibre rather than wood. The product is sold on the domestic market for a loss, and on the export market for an even larger loss. In view of this, it is difficult to understand why the Government has recently established another (wood based) particle board plant in the country. When this comes into production, the marketing problem will become even more severe. Moreover, the new plant will have other problems: it is to be based on the Chittagong Hill Tracts resource where, even for existing industries there, severe supply difficulties are already occurring.

3.3.3 Sawmilling

Sawmilling is a highly heterogeneous industry in Bangladesh, ranging from a fairly large Government sawmill, integrated with planing, preservation and other treatment units, through to itinerant pitsaws based in the rural area.

The large government mill is presently highly unprofitable, partly because of poor design and maintenance, and partly because no market appears to exist for its intended output of large sized material (for re-processing elsewhere), forcing it to base production on irregular, small-sized orders.

The small mechanical mills of the urban areas of the country, although besieged by the usual problems of poor power supply, irregular raw material availability and so on seem, by and large, to be able to provide the basic products required.

Little is known about the state of the rural sawmilling sector. In Douglas et al (1981), a residual estimate shows that it is a large sector. Since it has operated and continues to operate at a significant level without assistance or intervention from the Government, we must assume that it is reasonably profitable. However, from observation, it is an inefficient processor, in terms of sawn recovery, although the lack of preservation facilities must shorten the life of sawn products in the Bangladesh environment considerably.

3.3.4 Interpretations

There are a number of other industries in the Bangladesh forest products sector - match factories, furniture factories, packagers and so on - but the above group predominates, and serves to illustrate some basic factors about the industrialization route to forestry sector development in LDC's:

- there is a broad, inverse relationship between the size and complexity of established industries, and their relative

profitability. Production, resource and infrastructure problems
affect all industries in the forest sector, but the larger, more
capital intensive ones among them are least able to minimize their
problems themselves.

- basic demand problems also seem to affect the larger industries more
than the smaller scaled ones. Although the paper-making industries
are all now far too small to achieve economies of scale, by present
standards, they all produce far more than the domestic market can
absorb, but do not produce it at competitive international prices.
Consumption levels for the range of forest products is extremely
low:- this seems to lead analysts and planners alike into predictions
of substantial potential for market growth. In our view, however,
the time to invest in substantial capacity to serve the domestic
market is, when other conditions in the economy are such that real demand
is at a sufficiently high level.

3.4 Is There a Forest Industry Option?

The experience of Bangladesh with the larger scaled forest industries
- particularly the paper industries - has been essentially negative, and it is
an experience many other LDC's have shared as a result of the industry
optimism of the 1950's and 1960's. When these industries were established,
Bangladesh lacked (and still does) not only the infrastructural and
technological base to support them, but also the effective demand for the
products. This might perhaps have been bess damaging had the products
created found a ready export market but, almost inevitably when industries are
grafted onto undeveloped economies, the price and quality competitiveness of
the resulting output will normally preclude the export option.

However, it would be most unwise for us to extrapolate a general,
absolutist view on the matter from this example. Obviously enough, there are
countries in the LDC group where sufficient real demand for these sorts of
products exists to justify (technically, at any rate) some consideration of
the option to manufacture them - particularly given the fundamental role of

papers in education, literacy programmes and so on. In Bangladesh itself, there are some forest products which are highly in demand; which have an important export role (the plywood case); and which are manufactured there in a reasonably efficient manner. Such industries may not be optimal in terms of overall sectoral balances in the economy, but they at least do not make a negative contribution. Their operation gels reasonably well with the appropriate technology and basic needs yardsticks, as discussed in Chapter 2, and much of their output is indispensable to the basic functioning of the economy. The problem is really one of deciding which products to manufacture: in other words, as we have already generally remarked under subsection 2.3.2, the appropriateness of the product is as important as the appropriateness of its means of production. So far as the paper industry is concerned, an LDC government faced with a decision on this matter might do well to consider (in addition to the important distributional and socio-economic aspects involved) the following:

> "... the soundest and most reliable market for the justification of capital investment is a growing domestic economy in which paper products in the form of packaging or printing and writing grades have already begun to be established as an integral part of the economy of the nation."

> Gomez (1978)

Much attention has been given in recent years to the design of more appropriate processing technology for LDC's, particularly in the pulp and paper areas - these being generally the most problematical because of the scale and complexity of paper making as it has evolved in the industrialized world.

In 1976, FAO established the Pulp and Paper Industries Development Programme, which was aimed at producing workable designs for smaller scaled and simpler paper-making technology which would allow LDC's to produce reasonably cheaply. Leslie and Kyrklund (1980) in an assessment of this work, have drawn some relevant conclusions:

- the development of pulp and paper technology is now virtually confined to the industrialized world. Economies of scale of production in the West have become so large in paper making that direct transfer of the technology to LDC's is impossible.

- reasonably small scale units - capable of producing around 36000 tons of paper per annum-have been tested and shown to be technically efficient.

- mechanical pulping is also possible at an efficient level at fairly small scale. However, little progress has been made in the area of reducing the viable size of chemical pulping plants.

One interesting aspect of forest industry development which can best be examined by reference to examples is the matter of how the industry in question is designed and run. The first case concerns the establishment of a packaging, then package manufacturing enterprise in Colombia in the early 1950's. As explained by Gomez (1978), serious errors were made with this enterprise:

- the enterprise was, in a mood of general optimism brought about by high world demand for paper during the Korean war, extended into the paper manufacturing area, from the original concept of bag making from purchased papers. This development went not only beyond local expertise, but also beyond that of the United States based partner firm.

- the investment made was too large, in view of available funds, and installed capacity too great to serve the local market.

After the Korean war, this industry suffered very severe import competition from local packaging manufacturers who used cheaper, better quality imported paper. Had it not been for a general financial crisis forcing the Government of Colombia to implement broad import controls from 1956, it is unlikely the enterprise would have survived at all. In Gomez's view, although the enterprise prospers now, the decisions and risks taken in the 1950's were not justified, and are not vindicated by the fortunate long term outcome.

An observation on the method of operating the enterprise is relevant considerable success, apparently, has been obtained through a division of responsibilities between the partners. The US partner handles management training, finance and technology, while the Colombian counterpart oversights the legal, labour and social requirements.

A second paper making enterprise of interest here is one described Sila-On (1978): the Siam Kraft Paper company of Thailand.

This mill was designed as a quite sizeable enterprise, by LDC Standards: kraft paper and board capacity of 200 tonnes per day; a pulp mill capable of 60 tonnes per day. In all, the mill can meet 25% of Thailand's total paper requirements.

In common with many projects of this nature in the developing world Siam Kraft has had a less than impressive history. In its first 8 years of operation (through the 1970's) the company was on the verge of bankruptcy on three occasions. The Government eventually provided it with a monopoly on kraft paper manufacture in Thailand, and then instituted total import bans on competing imports.

Unlike many of the earlier exercises in LDC paper manufacture, Siam Kraft was, according to experts cited by Sila-On, equipped with well-built, good quality plant. The technology transferred was not the basic problem: financial structure and commercial management were. These have been approached in an unusual way: a large company, Siam Cement, took the operation over in 1974, and has been able to bring financial resources and management expertise to bear on some of the problems. Siam Cement hired a Japanese firm, Honshu Paper, to improve technological performance and know-how. This method of operating allows retention of local control over the enterprise (without stifling its operation with lack of funding), but allows immediate access to technological input and expertise from the Japanese firm. In this sense, it is an example of the approach to technology transfer suggested in Chapter 2 above - direct application of already available research and development expertise, rather than attempting to create technology within the LDC from a low base.

3.5   Forestry in Development: Newer Approaches

The lessons of experience of the industrialization approach to development have, in forestry, shifted the emphasis to other types of sectoral development.

Westoby himself, to his credit, has fully acknowledged (as we shall shortly see) that the approach to development of the forestry sector in poor countries advocated by him (in the article discussed above) and by others, has proved largely to be a failure. As we remarked in the opening sentences of Chapter 1, the practices and policies of this era are still persistent today in LDC's, even if the justification is not.

There is little point, we suggest, in attempting to re-construct the historical process by which the theoretical view of the role of forestry has undergone its radical change. According to the records of a working party meeting of IUFRO (in which the papers by MacGregor and Von Maydell cited above appear) in 1975, on the contribution of forestry to development, strong elements of the industrialization view were still apparent then, and the meeting seemed divided over its view of a paper by Haley and Smith (1976), which specifically criticized the use of indirect benefits, import replacement and disguised unemployment arguments in favour of forestry proposals. Haley and Smith in fact presented what might be termed the economic rationalist viewpoint; the primary purpose of investment in forestry, they remind, is not the growing of trees, but the formation of capital. It is very easy, they suggest, for planners and administrators to lose sight of this; to become overly attentive to the indirect, intangible benefits of what otherwise seems an inferior investment.

Westoby (1978), in his recantation of the views advocated in his earlier work, begins by referring to the inadequacy of the GNP criterion (see our discussion in Chapter 1). Westoby's new interpretation of the reasons for underdevelopment is interesting:

> "The underdeveloped countries are not underdeveloped because they started late in the development race. They are not underdeveloped because they lack resources. They are not

underdeveloped because they lack know-how. They are not underdeveloped because they are overpopulated. They are underdeveloped as a consequence of the development of the rich nations. The development of the latter is founded on the underdevelopment of the former, and is sustained by it. The ties between the affluent, industrialized countries and the backward, low - income countries are intimate and compelling. Their nature is such that the objective impact of most of the so-called development effort to date has been to promote underdevelopment."

This explanation, obviously, owes a great deal to the tenets of Dependency Theory (see our discussion in Chapter 2 above). Westoby argues that the growing interest in forestry projects by the development establishment in the 1960's had little to do with the idea that forestry had significant contribution to make to economic and social development, but was rather a result of the rich countries needs for raw material, and the opportunities offered to obtain it in underdeveloped areas. In support of this contention he cites the rise of export of tropical hardwoods from 3 million to over 40 million cubic metres between 1950 and 1976 - nearly all of which went to affluent, industrialized nations.

Westoby is pessimistic that mere recognition of the priority that should be given to forestry's role in the rural economies of LDC's by the large multilateral organizations and development banks will achieve real change. In his view, the principal problem is that governments and officialdom in many LDC's are simply not interested in mitigating the circumstances of the poor - that they are, in Dependency terms, part of the urban elite and determined to stay that way. In pursuit of this theme, Westoby claims that the demand for a new economic order in the world is a red herring, used by LDC governments to excuse their own shortcomings in the area of relief of the poor by reference to machinations of the rich countries against them.

Westoby's final peroration refers to the need for poor countries to concentrate on producing an agricultural surplus. The role of forestry in this must be to support the traditional rural sector, and in effect must be carried out _by_ rural people themselves. Industrialization must be directed towards satisfaction of basic domestic needs. Measures to reduce income inequalities will in themselves bring about change patterns of demand for forest products, but some intervention and manipulation of the market by government may be necessary (to direct and assist this process). Exporting has a role to play in the forestry sector, but must be subordinate to its domestic role.

Westoby suggests that forestry development along these lines might slow economic growth, as measured by the conventional criteria, but development, inclusive of the social aspect, will be accelerated. Discussing the forest industries that have been set up in LDC's in the past, Westoby observes that:

> "very, very few of the forest industries which have been established in the underdeveloped countries have made any contribution whatever to raising the welfare of the urban and rural masses, have in any way promoted socio-economic development. The fundamental reason is that those industries have been set up to earn a certain rate of profit, not to satisfy a range of basic, popular needs".

The Food and Agriculture Organization has also recently entered the field with a detailed policy statement on forestry in LDC's (FAO (1980)). This document also identifies the problem raised by Westoby: that few deliberate efforts have so far been made directly to improve the lot of the poor in LDC's through forestry activities. Forest services, operating on outmoded technical and societal perceptions, have been unable to effectively manage and control the resources under their jurisdiction. The general pressure of population today on forest lands, and the specific need of the poor to occupy and utilize such lands, and whatever grows on them, to sustain life, means that a new approach to management of forests (in densely populated LDC's in particular) is needed.

FAO identifies fuelwood, material for rural building and some of the minor forest food products as being high priority output. Integration of forestry with agriculture, voluntary participation of local communities in forest management and the use of appropriate technology are seen as the important elements in overall approach.

For governments of LDC's, FAO pre-supposes the political willingness to implement the necessary changes. Given this, the principal need is for creation and strengthening of national forest planning capabilities, forestry training and education, and community extension, to involve rural communities in forestry programmes.

For the FAO itself, the required role is perceived to be assistance to governments in planning a rural development strategy for forestry. FAO sees a need for its own operation to become more integrated, rather than separated into specialized areas, if it is to be effective in this.

The World Bank has issued an important statement of its policy on forest sector development (World Bank (1978)). This document brings the Bank's forestry sector lending criteria into line with the major re-orientation of Bank philosophy towards the concept of rural development which came about in the 1970's.

The Bank comments on the pronounced deforestation occurring in developing countries at the present time. Apart from the ecological consequences of over-exploitation and deforestation, the Bank identifies the serious consequences for rural welfare of increasing constraints on fuelwood availability, in particular.

The document provides a useful typology for forestry development projects, and it is relevant to briefly re-state these here to emphasize the point that there are many forestry problems, not one, in the various LDC's.

1. Wood deficit marginal lands: This type of forestry problem is exemplified in the Sahel regions : overgrazing and poor growing conditions have reduced existing vegetation and increased desertification. Without effective settlement, afforestation programmes would be extremely difficult to manage.

2. Potential afforestation areas: Similar to the wood deficient marginal lands, but with more favourable forest growing conditions and less population pressure.

3. Overpopulated wood deficient areas: Much of the sub-continent comes into this category, which is typified by overcutting of upland forests, with consequent erosion flooding and so on. Severe fuel shortages are already apparent, and institutional problems, with undermanned and poorly trained forest services, exacerbate conflicts with the local populace over forest use.

4. Wood abundant poor areas: Countries which are underdeveloped, but which have remaining a large, untapped resource of natural forests. The Congo, parts of Indonesia and New Guinea typify these areas. The major problem is to prevent conversion of forests to agriculture in such a way as to bring about rapid fertility losses in the fragile rainforest soils - a phenomenon which has been cited in many parts of the world.

5. Wood abundant areas with severe population pressures: Much of the world's remaining tropical forest is in areas under this category, in Africa and South America. Shifting cultivation poses the major threat to resources.

6. Wood abundant rich areas: Typified by Canada, the USSR and Scandinavia.

The Bank, in analysing its own role in forestry, emphasizes that its projected lending will be directed at rural forestry to a much greater extent than has been the case in the past : over half of its intended schemes will address this problem, compared to only a quarter or so of projects undertaken in the 1953 - 1976 period. Within this category, a significant part of Bank lending will be for projects of environmental protection - shelter belts, erosion control afforestation and so on, to preserve or accrete land for useful production by rural populations.

It now becomes necessary for us to take stock of the very different analyses of the forestry problem in LDC's that are implicit in these statements. It is a measure of the degree to which the general approach to development in very poor countries has penetrated forestry circles to note that Westoby, FAO and the World Bank do not differ in their basic assessment of what the major problem with forestry in LDC's is : a lack of involvement of, and benefit to, the rural poor in the mainstream of forestry sector activity, and a consequent aggravation of the problems of management of forested land and of the income gap constraint on development in these countries. However, the explanations offered as to why this situation has arisen, and consequently the suggestions advanced to correct it, are very different. Westoby's version of the problem is grimly political; the FAO view determinedly apolitical. The Bank seems more aligned to the FAO view, albeit that it does identify a tendency for LDC governments not to support forestry institutions to the extent necessary. We must attempt some general assessment of which view (if either) is close to reality.

Westoby begins, as we have seen, by suggesting in effect that poor countries are poor, because of the manner in which rich countries have made themselves rich. We have spent some time on this in our discussion of the New International Economic Order and Dependency Theory in Chapter 2, and our conclusion was that the trading disadvantages applying to LDC's through dealing with Centre countries do not appear to be as severe as some protagonists of NIEO claim. Nor, more prosaically, are they likely to alter significantly in the near future.

An important interpretation that Westoby makes for forestry, on the basis of his general explanation of underdevelopment, is his claim that the rich countries, running short of wood, have exploited the forest reserves of the poorer nations. This is probably so, but to what extent are continuing overall wood shortages a reality, and what advantages and disadvantages does the answer imply for LDC's?

Impending famines of wood in the world have in fact characterized forestry literature for many years and, to a large extent, still do so. Our view, which we have argued elsewhere, (Douglas et al (1977)) is that on current evidence, the hypothesis of an imminent, world wide and sustained shortage of wood is not well supported by the evidence available. It may be true, as is suggested by the World Bank (op cit) that something like 15 - 20 million hectares per annum of the estimated 1200 million hectares of mature forest in developing countries is being lost to over-exploitation, agricultural incursion and so on each year. This has serious environmental and economic implications for the LDC's involved, but what does it mean in terms of international supply? Many developed countries or regions of the world are now following policies of raw material self - sufficiency. Even Japan, the quintessential raw material importer, now has a very considerable 9.4 million hectares (see Byron (1979)) of man - made forests, (much of it currently in an immature state). Matsui (1980) gives some growing stock data for Japan which imply a standing volume in that country of $20m^3$ per capita - quite an appreciable amount.

It is interesting to note that the Food and Agriculture Organization - previously amongst the most pessimistic of forecasters in terms of supply shortfalls - now acknowledges that the world supply of industrial wood will meet additional demand to 2000 AD (the forecast period) (FAO, 1981).

Wood is a renewable resource. In fast - growing plantations, rather small establishments can replace quite large areas of natural forest, (so far as wood volume production is concerned, if not for ecological, recreational and other purposes). Some developing countries may well run short of wood themselves, but in our view it would be unrealistic of those of them with surpluses to expect massive trade advantages to come their way for their remaining resource, because of overall shortages in the world. It is undoubtedly true that relative to their own histories of absolute abundance of wood, some rich nations have recently experienced greater need for forest raw material from elsewhere, and have engaged in highly exploitive operations to get it. But, because of the renewability of the resource, and existing re-afforestation policies in the West, this sort of shortage is quantitatively and qualitatively different from that for, say, oil. Wood exporting LDC's would, in our view, be most unwise to expect dramatic results even if they do form consortia or cartels to control supply. The fact that oil - rich countries were able to do so, and simultaneously to overcome a number of trading disadvantages (the presence of multi-national, Western based extractive industries; a history of colonial exploitation; concerted pressure from the Western trading block against the cartel) should not be taken as a precedent for the wood exporters, albeit that Westoby and the World Bank occasionally seem to come fairly close to suggesting this. Clark (1974) provides an interesting analysis of the characteristic price response of wood in its markets. He shows that a very definitely convex demand function for wood applies. This indicates that the commodity is one which finds a wide range of uses when cheap, but a wide range of substitutes when expensive. whilst we read much about the high income elasticity of demand for wood products, we seem to see rather less on this demand/price response in the literature on overall supply and demand in the international wood markets. It indicates that demand will adjust to supply well in advance of serious sustained shortages in this market, and the possibility of offsetting supply responses in deficit areas is also, as already discussed, an important factor. Prices for the raw material may rise on the international market but, despite what is indicated by rather simplistic static supply/demand analyses, we do not believe that wood - surplus LDC's should expect real pri rises to be large, nor should less well endowed LDC's enter the plantation supply market on the assumption of radical price rises. Better control of natural resources in wood rich LDC's would undoubtedly be to their economic advantage, and there are good theoretical and practical reasons for LDC's to trade in forest products with Centre countries to mutual advantage. Our argument is merely that this should be kept in some perspective. Wood shor LDC's, in particular, should pay most attention to their own specific needs when determining resources policy.

In fact, Westoby himself seems ambivalent on the question of exploitation of LDC forest resources. Later in his paper, he suggests that the issue of the New International Economic Order is irrelevant. Now, eith the Centre countries have been exploiting LDC's as Westoby first suggests, i which case the latters' demands for a new international trade and aid arrangement are justified (no matter who makes them) or they have not (in which case Westoby's accusation of LDC governments of using such claims to mask exploitative tendencies of an internal origin in such countries becomes relevant). But both cannot apply.

Westoby's interpretation of what has occurred in the past so far as the establishment of forest industries in LDC's is concerned also seems faulty, albeit that his overall conclusion (that these industries have not made any significant contribution to development), although drawn for the wrong reasons, is more tenable. Westoby says that the reason for this is that these industries have been set up to earn a profit. He seems to have this same point in mind when he suggests that forestry development along the lines he suggests might slow the rate of economic growth. We would dispute both observations. We have seen, in Chapter 2, some overall results on income growth and development: from these analyses, for very poor countries we would immediately reject any proposal which threatens to slow growth but, as Lipton has shown, this should certainly not exclude programmes based on rural development (nor, by extension, forestry schemes within such programmes). And the reason the previously established forest industries in LDC's have proved so detrimental to socio-economic development is not so much that they have been set up to make profits, but rather that they have usually failed to make them. The basic reason for this, as we have already suggested, is because they were not appropriate industries in the first place. Had they made substantial profits, then we suggest their reputation and standing today, and the current view of development overall, would look rather different. Redistribution towards impoverished groups is possible (if not always practiced) when industries are making large profits - unassisted by government subsidies of one sort or another. It is impossible when they are operating at a loss.

It is rather more difficult to critically analyse the FAO statement on forestry in LDC's, largely because it is a somewhat bland and amorphous document. It offers little direction on implementation of its recommended approach, and even less on how the trade - offs it implies should be managed. That there will be trade-offs in many LDC's should be obvious : they have in the past established (very often on the advice of FAO and other large international assistance agencies) forest growing and processing sectors which, they are now being advised, are inappropriate for their development aspirations. We will have more to say on this shortly. The document speaks of strengthening planning capabilities, broadening the administrative bases of forestry: and so on. Whilst these appear to be positive exhortations, we will argue below that their consequences might be anything but positive. They certainly offer no direction as to how dis-investment of existing industries should proceed, nor do they address the fundamental problem, of incorrect or inappropriate perceptions of the role of forestry.

In any event, the document appears somewhat in two minds on the question of industry policy. On the one hand, FAO is critical of the establishment of forest industries in the past, for reasons of the lack of involvement of the poor in the sector. Then, later, it is asserted that there is no need to constrain the poor to simple and small technology. To do so, it is claimed, raises the possibility of unacceptable dualism in the development pattern. FAO suggests that appropriately designed technology, capital intensive and sophisticated, can be used to directly benefit the poor. Given the absolute shortage of investment capital, the manifest difficulties in establishing capital intensive industries (and demand for the products) in LDC's, and the overwhelming need to employ humans, rather than machines in these countries, we would have to disagree. Of all organizations, FAO should be well aware that the justification for the capital intensive, technological approach to development in the past has always been that, eventually, the poor would benefit. And all the resulting industry that was set up as a result of former theories on development was, at the time, believed to be appropriate and efficient. FAO's argument here would seem to be a reversion, mid-stream, to the older concepts of development. It is, furthermore, nonsensical to suggest that small and simple technology is "socially unacceptable" in the sense of promoting dualism. The alternative for the poor, as ought by now to be abundantly clear, is not meaningful involvement in the so - called modern sector of the economy, but rather, little or no involvement in anything at all - surely a much more extreme strain of 'dualism'.

The World Bank document also seems, on occasion, to drift into this type of reasoning. The Bank suggests, at one point, that LDC's have a comparative advantage in the growing of wood, insofar as their better ecological conditions and cheaper inputs allow this. Wood, however, is expensive to transport in the raw form (not sufficiently so, it would seem on the Bank's own arguments, to have prevented large scale external commercial exploitation of forests in the past). The Bank's solution is primary and secondary processing within LDC's. We shall shortly come to a reasonably successful application of that principle, but our review of the often disappointing performance of large scale forest processing industries earlier in this Chapter should serve to warn of the dangers of too much enthusiasm allowing highly technical, inappropriate processing tails to set about wagging the resources dog all over again.

3.6   The Malaysian Example(2)

Malaysia is a case of a developing country where the forestry sector has made significant contributions in both developmental and distributional terms and for this reason, it warrants some attention here.

Malaysia has some distinct advantages: a relatively small population (12 million or so) for its land area, and abundant resources. On Peninsular Malaysia, only about 20% of land area is currently farmed; in Sabah and Sarawak, this figure is only 5%. The remaining area of the country is forested with a resource of variable quality and quantity.

On Peninsular Malaysia, forest utilisation and management has been intimately connected with both the development of a fairly sophisticated and competitive export processing sector, and rural development. In 1981, about 4.4 million m^3 of logs - about 55% of total harvests on Peninsular Malaysia - came from clearing operatons for agriculture, and the Government of Malaysia quite obviously sees considerable socio-economic benefit in this sort of operation.

However, of the total log harvest on Peninsular Malaysia, les than 3% is exported in log form. The remainder goes into locally established sawmilling and ply operations, predominantly aimed at the export market. These exports earned Malaysia $US684 million in 1980. At present, integration of secondary and tertiary processing is being developed.

A problem is beginning to arise, in that developments in processing in Peninsular Malaysia have outpaced log availability. The Government is reducing the rate of harvesting, and is attempting to introduce improved management and utilisation. Also, what seems to be a highly optimistic plantation programme to establish 188000 ha per annum for the 15 year from 1981 of exotic conifer plantation, has been announced.

In Sabah and Sarawak, the emphasis is still on export of log material: 94% of forest exports from Sabah and 82% from Sarawak, were in log form in 1981 (total forest exports from these states totalled $US 1324 milli[on] in 1981). Presently, Sabah is trying to lower the log component of exports somewhat by creating processing facilities there.

Of the 416 000 tonnes of paper consumed in Malaysia in 1980, only 6[3] 000 tonnes was produced locally. Recently, plans to use the softwood plantations on Peninsular Malaysia for paper making were shelved, so that the resource can go to existing wood processors. Plans exist for construction [of] a 200 000 tonne per annum pulp operation on Sabah: even with construction of this facility, Malaysia would remain dependent on imports for the bulk of its paper requirements.

The Malaysian forestry sector, on the basis of the above brief sketch, demonstrates a number of interesting features:

- a successful export industry has been established on Peninsular Malaysia on the basis of the relatively low level technology required in sawmilling and veneer production.

- the Government of Malaysia has control of forest clearing and allocates land to rural dwellers for agricultural development from the process. Even given the impending resource problems on Peninsular Malaysia, this integration of forestry and agricultural responsibilities has arguably produced a more orderly and equitable distribution of wealth than might otherwise have occurred.

- although a relatively efficient processor, Malaysia as a whole still generates more export income from log exports than from processed forest output. The question of how quickly the transformation to domestic processing in Sabah and Sarawak should be made remains contentious.

- Malaysia has so far not made significant attempts to replace imports in the paper sector. Whilst pressures in some quarter to do so are mounting, the Government needs to bear in mind that the nation is located close to large paper producers in the South East Asian region.

Without attempting to minimize some of the basic forestry problems faced by Malaysia, it seems reasonable to interpret the overall performance of the sector in a favourable light: control over forest exploitation has been maintained; a rural development/redistribution element has been included in forest operations; specialization of industry activities into areas of comparative advantage (even to the extent of retaining a fairly large raw material component in exports) has established Malaysia in some important markets; and justifiable caution has (so far) been applied to the question of investment in high technology, capital intensive production processes such as paper-making.

Chapter 4
FORESTS, FUELWOOD AND PEOPLE : THE LINKAGE

As we have seen, the emphasis in international forestry development circles has now shifted towards a recognition of the need to link up forestry development with rural development - at least, in those LDC's where crowding and/or competition for basic productive resources is severe. This change in emphasis does not, as we shall elaborate upon in the final Chapter, obviate th need to rectify, adjust and impove the processing sectors which have been set up in many LDC's, albeit that a reading of some recent literature on the subject might suggest that that problem is no longer of interest.

As implied in the title to this chapter, there are strong causal relationships running between the condition of forests and forest lands in LDC's, and the socio-economic situation that pertains in their rural populations. It needs to be said at the outset, however, that the apparent uniformity of the current problem - deforestation in an uncontrolled manner, in the company of growing masess of rural poor - should not tempt us into too many generalizations about solutions. One thing we will attempt to show in the remainder of this book, is that the project planner and the development theorist alike must understand the sociological and cultural causes of poverty, as well as the macroeconomic parameters of it, if their prescriptions to alleviate it are to succeed. These sociocultural factors ca be highly variable, between countries and even regions within countries, and appropriate measures to produce sustainable changes in the welfare of individuals will need to be similarly variable.

Although the relationships between environmental, social and economi factors in LDC forestry are many, and strong, we need to partition the discussion of them somewhat, to impart some comprehensible structure to the argument. In what follows, we will move from a very brief review of the overall role of the tropical forests in the environment, to a consideration o what changes are currently occurring in these forested areas as a result of man's intervention. The most important single product from LDC forests, from both the environmental viewpoint and, at the other end of the scale, the basic economic welfare position of most of the people, is fuelwood, and this warrants some separate discussion below.

The chapter will close with a review of some rural forestry projects.

4.1 Tropical Forests and the Environment

4.1.1 The size and influence of the forest

More than half of the land area of tropical countries - around 1200 million hectares in total are still covered by humid forests. If we include the sparse woodlands and the shrubbed areas of the more dry areas, then a total area of about 2500 million hectares remains unaffected by agriculture at the present time.

As we will see, a great deal of tropical forest cover is being removed, one way or another, at the present time, and some of the influences of forests on the ecology are therefore relevant. According to Fontaine (1981), the predominant characteristics of tropical soils is a high degree of alteration of parent rock: the result is poor mineral reserves, poor natural nutrition and low absorptive capacity. The enormous biomass production of tropical sols is, in some senses, an illusory phenomenon: it comes about through highly efficient re-cycling of organic materials from the growing matter through the upper soil layers. Although regeneration of forest cover on soils of this nature is possible, it is by no means a simple task once the soil has been cleared of natural vegetation and exposed to leaching and erosion of the thin nutritious layer. Shifting cultivation, when too intensive or frequent, and fixed agriculture can make it impossible for tropical forest soils to regain their fertility. Water values downstream from the cleared areas can be affected - turbidity and rates of flow are altered, erosion, flooding and shortages can be the result. The World Bank (1978), for example, attributes increased flood levels in Pakistan in recent years to siltation of the Indus river system, brought about by excessive forest cover removed in higher catchment areas.

Less well documented is the effect of forest removal on microclimates, and perhaps of thermal balance at the global level. A good deal of controversy surrounds this subject and we can do no more at this point than reiterate Fontaine's (1981) suggestion that what is known counsels caution in exploitation of tropical forests.

The following table, adapted from Fontaine, shows the gross areas of forests remaining in the tropics:

Table 4.1

World Areas of Tropical Woody Vegetation ('000 ha)

	Productive closed forest	Unproductive closed forest	Open forest	Total
Tropical Africa	163 050	53 600	486 450	703 100
Tropical America	521 680	157 000	157 000	836 650
Tropical Asia	201 000	104 500	30 950	336 450
Total	885 700	314 100	674 400	1 875 200

To these areas can be added a further 402 million ha for fallow forest areas, and another 624 million for shrub formation areas.

Deforestation in closed forest areas has been estimated, and is shown in Table 4.2 below:

Table 4.2

Deforestation in Closed Forest Areas ('000 ha)

	Productive closed forest	Unproductive closed forest	Total closed forest
Tropical Africa	1 262	71	1 333
Tropical America	3 049	1 070	4 119
Tropical Asia	1 697	118	1 815
Total	6 008	1 259	7 267

Overall, then, deforestation in recent years in the tropical zone has been running at about 7.3 million hectares per annum, i.e. about 0.6% per annum of the total resource. This figure is higher than FAO's (1980) estimate of 5 million hectares cleared for new replacement (presumably because of the shifting agriculture factor) and considerably lower than an estimate of 15-20 million hectares per annum by the World Bank (1978).

4.1.2 The Use of the Forest

Whilst, as we shall see, the pressures on forested land form are from one form of agriculture or another are significant, there is also the factor of the difficulty of sustained yield management of the tropical forests to be taken into account. Tropical forests are often in remote inaccessible areas, which discourages intensive management. Also, a large proportion of the resources occurring are presently non commercial and there is great variability in the stocking of tropical forests, ranging from around 60 m^3/ha in the dry dense forests of Africa, up to 270 m^3/ha for the semi-deciduous forests of the same continent. Intensifying shift and burn cultivation in forest areas, giving cutover areas insufficient time to recover before the cycle begins again further complicates management. As Leslie (undated) has remarked:

"... . after a century or so of determined effort to mould the tropical moist forests into a managed natural state the solution to the problem is apparently to be found in abandoning the natural forest".

Leslie goes on to warn that even though economic calculations may seem to support this idea, it may be based on too narrow a definition of the benefits of sustained forest cover in such areas, omitting important social and economic variables.

Leslie, among others, claims that sustained yield management of tropical forest areas is possible. However, it is fair to say that many LDC governments do not see it so, and have left their forest resources to uncontrolled commercial exploitation (which has been significant, for example in the Philippines and Brazil); to the activities of the shifting cultivators; and to encroachment of fixed agriculture systems. The FAO (1980) notes that, about 75% of the land clearing in LDC's is unplanned (by governments). This, it should be noted, does not necessarily mean it is undesirable.

Agricultural production is, in many instances, a higher priority goal than forestry, and the trading off of certain water and soil values under forest cover for agricultural production may well be a valid transformation - as indeed it is in Malaysia. However, rational decision-making is rarely possible in this area - the information on what effects changed land use practices are likely to produce is normally not available to governments, even if they were in a position to control clearing and settlement. And it is fairly clear that a lot of clearing of this type has had serious consequences for land productivity, forest options and, ultimately, the welfare of the rural populations dependent on the resource.

There are two perspective points to bear in mind, when considering information of this nature, on a world wide basis. The first is that in many places in the world - particularly in Africa - the major factor in deforestation is wood-gathering by local dwellers (principally for fuelwood) rather than clearing for agriculture.

The second point is tht the large areas of naturally occurring forests in LDC's are not the whole of the forest sector. Some countries already have significant large scale plantation resources. Others have large resources in village, homestead and communal forests. Even in Indonesia, for example, which is a large forestry country, it is obvious that the bulk of the population - on densely populated Java - would regard the planted (or regenerated) village resources and agroforestry resources as of far greater importance than the large natural forests of the outlying islands. And in Bangladesh, the planted, privately owned homestead forests supply far more wood volume in toto than all of the country's natural forests and large scale plantation areas combined. In the well-known Chinese example, too, the fact is that the 28 million hectares of land in agricultural areas that has been planted with trees since 1949 is now the dominant source of wood fibre for the majority of the population.

Whether an LDC is in a condition of rapid forest depletion or equilibrium, or actual increase in its forest resource, the predominating reasons will be found in the sociological factors governing the relationship between rural dwellers, land use, and the forest itself. This may seem an obvious conclusion, but it is one that has been particularly slowly recognised by forest sector administrators (and their advisors) in the developing world. Even now, as we saw in Chapter 3, it is possible to find lapses in logic in published strategy documents on this subject from the large international aid agencies, and it is certainly true that relations between forest services and rural dwellers in most of the countries in the World Banks 'overpopulated wood deficient' category remain extremely poor. And the results of this situation, for the environment and for rural welfare, remain damaging.

Bangladesh provides an example of a forest depletion situation which is typical of many parts of Asia. Running down the eastern side of the country is a zone of low hills, known as the Chittagong Hill Tracts. Originally, virtually the entire area of the hill tracts was occupied by mixed tropical hardwood forests, with many species of the 'Malayan' type present: Artocarpus chaplasha, Swintonia floribunda, Gmelina aborea, various Amoara and

Dipterocarp species, and many others. Bamboo in natural stands was also widespread in this area.

Of the 13 000 or so square kilometres of the Chittagong Hill Tracts, over half is now in a deforested condition, and because of the steep slopes and relatively high precipitation involved, significant erosion has occurred, causing siltation problems in a hydroelectric project in the area and in irrigation schemes on the south-eastern parts of the delta plains below. The circumstances tht have led to this situation, are as follows:

1. The area has been the traditional home of tribal groups - the Chakmais being the largest group - who currently number about half a million. These people are ethnically and culturally different from the Bengalis of the plain area.

2. The principal means of livelihood of the tribal groups has always been slash-and-burn agriculture in the forest areas.

3. The tribal groups have been under considerable pressure for many years from a number of sources:

    - in the 1960's, a hydro-electric scheme was constructed on the Karnaphuli River. The resulting lake submerged a large area of tribal land.

    - in the 1970's, during the period of Awami League government, large numbers of Bengalis were moved into the area, in an attempt to establish permanent agriculture in the denuded areas

    - the hill tracts is the site not only of the major remaining natural forests of Bangladesh, but also of plantations of teak, which have been established there for many years. Relations between the forest service and the tribal people have been hostile: the forest service, by and large, operated on a philosophy of complete exclusion of the tribal people from the hill forests for many years.

There are now some changes in attitude taking place, on the part of both Bangladesh Government officials, and the tribal groups. Some experimentation with agro-forestry and other resettlement schemes is taking place. However, at this stage it must be admitted that the problem is proceeding faster than the solution. A considerable military presence is now established in the Hill Tracts, and outbreaks of fighting are common. The severity of forest and other operations in the area is under constant risk. Poorly conceived projects still are initiated, and exacerbate the situation. In one, a large road was planned to join two major centres in the hills - but little consultation with the tribal people was undertaken, with the result that they suspected (perhaps with some justification) that the principal purpose of the road was to improve access for the Bangladesh Army. Considerable disruption and sabotage of the project ensued, and recently it was abandoned.

The situation has now deteriorated to such an extent that it is difficult to conceive of realistic solutions. Given time, and a reasonable degree of tact and generosity on the part of the Government (and the agencies which attempt to assist it in this matter), it may be possible to settle the tribal people into permanent agriculture, or agro-forestry - as is being done in northern Thailand with the Meo tribal people - apparently with some success. But, even if this is achieved, the environmental damage set in train by the historical situation will continue. It is simply not feasible, in our view, for a country such as Bangladesh to consider the re-plantation of six or seven thousand square kilometres of land - or even of enough of it to mitigate the major erosion effects (unless quite massive aid to do so is forthcoming from somewhere). Unless some low cost method of holding the soil (perhaps aerial seeding with selected grasses) can be implemented as an interim measure, it would seem that a large part of the Hill Tracts is destined to remain degraded and unproductive for the forseeable future. Certainly, if a forestry option is considered at all, it will need to be based on the use of pioneer species (suitable usually for fuelwood purposes, rather than industrial use), which are adapted to degraded sites.

Similar problems of intensifying shift and burn agriculture by hill tribes, leading to deforestation, are occurring throughout the South East Asian region. Few, however, yet face the serious overcrowding situation of Bangladesh, nor the degree of environmental damage already done. In Bangladesh, any projects aimed at reafforestation, agroforestry or purely agricultural settlement of the hill tribes will need to spend their early phases researching the social, political and economic background of the problem and testing alternative approaches to solving it. Otherwise the ris of ineffective, or even counterproductive activity is very high.

4.2   The Fuelwood Question

4.2.1 The general situation

The crucial underlying question which must be posed, when considerin the problem of conserving or creating forest resources in the LDC environment is: what is the primary goal of forest output? In the past, as we have seen the temptation was to answer that the purpose of forest raw material was primarily to drive the engine of industrial development. In wood and energy abundant LDC's, this is still a reasonable approach for government, on the assumption that rural dwellers will have sufficient access to forests to satisfy their own priorities. But such countries, we suggest, are fairly rare in the LDC group. Most LDC's have an energy problem which will have to be satisfied at least in part on the basis of domestic resources and many hav already begun to experience shortages of one such resource: fuelwood.

We cannot, in this book, do much justice to the large technical, economic and social issues that underlie the fuelwood question: the intereste reader is referred to some recent World Bank or FAO publications on energy an fuelwood (World Bank (1979), (1980), FAO (1981)), and to Earl (1975) for some technical background material. One thing is clear from recent studies: most of the population of the world now lives in areas where fuelwood is important and is also either in deficit, or about to become so.

Table 4.3 below reproduces estimates and projections of populations involved in fuelwood deficit situations from FAO (1981).

Table 4.3
Fuelwood Deficit Populations (millions)

Region	1980						2000	
	Acute Scarcity		Deficit		Projected Deficit		Acute Scarcity or Deficit	
	Total	Rural	Total	Rural	Total	Rural	Total	Rural
Africa	55	49	146	131	112	102	535	464
New East & North Africa			104	69			268	158
Asia/Pacific	31	29	832	710	161	148	1671	1434
Latin America	26	18	201	143	50	30	512	342
	112	86	1273	1052	323	280	2986	2398

The World Bank (1978) has estimated that, even with some fairly optimistic assumptions about the introduction of appropriate energy technology in LDC's (more efficient stoves, biogas reactors, solar cookers) an additional 20-25 million hectares of forests for fuelwood purposes will be needed by 2000 AD. This considerably exceeds the level which will be achieved by present rates of planting for this purpose. In some areas, such as the well known example of the Sahelian zone of Africa, the problem will be far more acute.

In overall terms more than half of official estimated wood usage in the world goes to fuelwood (or charcoal). In 1975, according to the World Bank (1978), around 550 million m^3 of industrial wood, and a further 133 million tonnes of paper and paperboard products were consumed. It is unlikely that in terms of roundwood equivalent, this would exceed the fuelwood and charcoal consumption estimate for the same year of about 1200 million tonnes, and, so far as LDC's are concerned, this makes fuelwood volumes of consumption far more significant, since something over 90% of all fuelwood is consumed in LDC's, whereas less than 15% of industrial forest products is consumed by them. Moreover, official estimates of fuelwood consumption are likely to understate real usage levels by a considerable amount in many LDC's: much of the removal of fuelwood from government forests is illicit, and does not enter official estimates, and very few adequate consumption studies for fuelwood have been done.

We are not, in this book, concerned with making out an unassailable case for development of fuelwood reserves in LDC's as compared, say, to the introduction of energy-saving technology at the rural level, or the various alternative means of energy creation: solar;wind; biogas; or even (in some areas) more develpment and reticulation of natural gas or hydroelectric power. We can safely assume for our purposes, that heavy reliance in many LDC's on fuelwood is going to continue for the forseeable future. Moreover, on the bases of current trends in removal and/or usage of natural forests or planted village/homestead resources, it seems clear that foresters and rural dwellers alike are going to have to become more engaged in fuelwood resource creation and conservation than has previously been the case: it needs to be borne in mind that in Africa, for example, virtually no creation or replacement of fuelwood resources has ever occurred: it has simply been gathered from the bush. We need no longer go to the well known case of the Sahel zone to find places in Africa where the distances which have to be travelled to gather wood fuel have become so great that one or even more members of a family will be fully occupied on this task alone. The capital of Mali, Bamato, used once to be completely supplied with its fuelwood needs from a radius of less than 50 kilometres. At present, the distance is over 100 kilometres and the World Bank (1978) argues that by 1990, Bamako will need access to a plantation resource of over 100 000 hectares to supply its needs – well beyond the capacity of the nation to establish.

Wood as an energy source does have some characteristics which ought to make it a reasonably attractive proposition for LDC's concerned with rural development, basic needs and more efficient use of productive resources.

- it is a renewable, and creatable resource: it can be established in areas where it has not been previously available (in Ghana, for example, Indian neem trees were introduced on the Accra plains, and have provided fuel ever since. In the Philippines, the fast-growing legume species Leucaena leucocephala (Ipil-ipil) has been planted in areas where the noxious weed Imperata had occupied sites, and the plantations have since been producing volumes of fuelwood in excess of 20 m^3/ha/annum).

- it can be grown in situ, obviating the need for transport systems, and using simple, locally available techniques. On the other hand, it is apparently feasible to use woodfuel in more sophisticated power generators: the Philippines has plans to use Ipil-ipil in this way, to fuel electrical power plants.

- according to the US National Academy of Sciences Research Council (in an article in Development Digest, Vol. XIX, No.3, July 1981), about 75% of tropical land is unsuited to pure agriculture, but about half of it would support tree crops - in combination with underplantings of annual crops or grazing. Many species of trees suitable for use as fuelwood are also 'pioneer' species, well-adapted to degraded soils, wind and droughting effects, and often suitable for colonization of deforested areas.

- the silvicultural technology of fuelwood growing, although not particularly well researched in comparison to industrial forestry, is inherently simple: the size, shape and branching pattern of the product is relatively unimportant, so long as the species grows fairly fast, and produces combustible biomass. Enough is already known for successful fuelwood plantation establishment to have fairly good prospects of success in most places - provided that the plantings can be protected and maintained. Moreover many fuelwood species are of the coppicing type, which obviates the need for re-planting after harvesting.

- some fuelwood species can have important by products in certain areas: beneficial erosion control and site quality effects; shelter; provision of food, fodder, mulching material and so on

Given these sorts of advantages (and the list could be easily extended), it may be wondered why the so called 'second energy crisis' - the shortage of traditional rural fuels in LDC's has developed into such a seriou problem. There will, of course, be a great many country-specific reasons, but some general observations seem relevant:

1. As we have already seen in our general analyses in Chapter 1 an 2, many LDC governments, and development assistance agencies, have been slow to recognise the importance of the rural sector, and the seriousness of its problems. In many countries (we will examine a case in point below), governments simply did not recognise that traditional fuels were becoming constrained.

2. The major concern of forest services in many LDC's has been the management of large scale forest resources, principally for production of industrial raw material. To be realistic, many of them find it a considerable strain on their resources and expertise to manage this task, and so the suggestion that they should extend their activities, into an area which is well outside their traditional forestry training and perceptions, is rarely given a sympathetic hearing.

4.2.2 Fuelwood and the rural poor in Bangladesh

We can learn more, at this stage, by examining the situation that pertains in a specific country. Our discussion here will draw heavily upon results and conclusions reorted in a paper dealing with the traditional fuel problem in general in rural Bangladesh (Douglas (1982)).

At the time the research reported in that paper began in 1979, it was not possible to say whether or not a shortage of traditional fuels existed in Bangladesh: this in fact is a quite difficult question to answer, because in Bangladesh the mix of products used for cooking purposes in the rural areas is complex - ranging from tree fuels, through the great number of combustible agricultural residues (including animal dung, which is an important fuel throughout South Asia), to such low grade fuels as dried water hyacinth and the like. It is simply not feasible to measure the supply of all these products in the rural environment - but if we cannot do this, how do we determine whether or not real fuel shortages are a problem? One answer is to use an indicator, which can be measured, and in Bangladesh it was possible to use the homestead forests in this way, because data on this resource was becoming available.

Trees have always been planted by rural dwellers in Bangladesh, as sources of fuel, fruit and nuts, fodder, structural material and a wide range of other products. Plantings are done on homestead mounds, by individuals - there are no communal or 'village' forests. Virtually all trees in the rural area of the country are there as a result of human settlement: they have been planted, or at least encouraged to regenerate. Traditionally, village dwellers of Bangladesh have been well disposed towards tree husbandry, and well aware of the techniques involved.

Under such circumstances, the condition of the homestead forests can be used as an indicator of the overall fuel situation. We argue this on the following bases:

- trees are an important fuel source;

- trees are, in a sense, a fuel capital stock: unlike the other traditional fuels, more can be removed in a given period than has grown in that period, simply by raising the felling rate;

- given the long term consequences of doing this, the fact that trees are private property in rural Bangladesh, and the observations that the consequences of over-felling, as it is known, would be known to rural dwellers, over-felling can be taken to indicate that insufficient other fuels are obtainble - i.e. that a general shortage of fuel exists.

The above may seem to be labouring the point somewhat, but it is only since results from recent surveys and subsequent analyses based on this sort of reasoning have been available that general acceptance of the seriousness of the rural fuel situation in Bangladesh has been accepted. Until that point, some results from a large energy study done in Bangladesh is 1975 by consultants hired jointly by the Asian Development Bank and the United Nations Development Programme (Montreal Engineering et al (1976)) had been accepted. That study had suggested, on the basis of some rather cursory homestead forest estimates based on aerial photography, that the homestead resouce was in equilibrium. This led to a low estimate of per capita fuelwood usage (and energy usage overall) - so low, in fact, as to be virtually impossible. Nevertheless, it was accepted, and led to a certain complacency about the condition of the resource.

Recent detailed survey results on both consumption and standing volume in rural Bangladesh are described in Douglas (1981), and Hammermaster (1981). Interpreting and combining these results (see Douglas (1982)) leads to the finding that, at the time of the surveys (1979/80) the homestead resource was being felled at an annual rate of 10% of standing volume. Even allowing for some error, and for the possibility of an anomalous rate of felling in that period, this is an alarming result. The homestead resource, by its nature and composition, is incapable of growing at a rate of more than 4-5% per annum, at the most.

Thus, the general Bangladesh Energy Study finding in 1975 that the homestead forest resource was in equilibrium is clearly a serious error: the resource is now, and certainly must have been then, in a state of rapid depletion. In passing, it should be pointed out here that there is great danger involved in publishing hastily compiled 'ball-park' estimates on matters such as these. No matter how circumspectly the interpretation may be made in the publication itself, it is the basic figures which will influence decision-makers.

The implication which can be drawn on the basis of using the homestead resource as an indication, is that fuel shortages are now a serious problem in rural Bangladesh.

Whatever else, this is certainly not brought about by excessive consumption of traditional fuels. Table 4.4 below reproduces some results from Douglas (1982) on traditional fuel consumption:

Table 4.4

Consumption of traditional fuels in rural Bangladesh (per capita per annum)

Household Land Area (ha)	Fuelwood (m^3)	Other tree Fuel (kg)	Bamboo (kg)	Agricultural Residues (kg)
0 - 0.4	0.072	165	25	178
0.4 - 0.8	0.073	172	32	149
0.8 - 1.2	0.072	126	34	149
1.2 - 1.6	0.071	110	29	132
1.6 - 2.0	0.077	164	27	125
2.0 - 2.6	0.090	112	31	111
2.4 - 2.8	0.078	87	27	126
2.8 +	0.109	96	23	75
Weighted Av.	0.079	131	28	143

These figures, when converted to energy equivalents, suggest that the average consumption of energy in rural Bangladesh is around 1.1 million kilocalories per capita per annum - an extremely low figure.

It is interesting that the figures also indicate little rising trend (in total energy usage) with income (which in this context, can be assumed as fairly closely correlated to land holding). This is a counterintuitive result, and is at variance with cross-sectional energy study results (see, for example, Earl (1975), Hrabovsky (1980)) across wider income ranges which demonstrate rising energy use with income. It is certainly a fact that wealthier families in Bangladesh live in larger groups, and some efficiency effect in food preparation could be one result. Also, the mix of fuels differs across income strata, and the median heat efficiency of the fuels used by wealthier groups may exceed that of poorer groups, even if technically measured kilocalorific content does not.

A more intriguing observation, from the point of view of our major interests in this book, is the relatively constant use of fuelwood amongst all groups (except the wealthiest stratum). There are two important points to consider here:

1. In Bangladesh fuelwood is most certainly not a free good. It is either retained by its owners for their own use, or sold at prices which, at the time of writing, were extremely high (Tk 550 - 700/m^3).

2. Poorer rural dwellers in Bangladesh do not have trees of their own - or at least, have very few.

The clear implication is that, in rural Bangladesh, the rural poor are being forced into the market place to purchase fuelwood. This was not always the case. Briscoe (1979) points out that historically in the Bengal area, the various fuels were in relative abundance and the rural poor were traditionally afforded the rights to forage for residue fuels on wealthier farmers' lands.

When energy from traditional sources becomes scarce, a vicious cycle ensues: agricultural residues that once were used to feed animals (an important form of power for necessary deep ploughing of the soils), or directly as organic enrichment of the soil, are increasingly diverted to fuel purposes: and soil fertility drops as a result. In the case of Bangladesh we can see that the largest proportion of the economic burden of fuel shortages will fall on the poor, landless group. We see, therefore, in this example the strong relationship betwee fuel and food production in the rural environment, and the particular distributional and welfare effects of fuel shortages. The resulting pattern is more or less as we would expect, given our general observations on the socio-economic structure of rural Bangladesh in Chapter 2 of this book: as fuel becomes shorter, poorer groups in society will be forced to spend greater proportions of their income on it. The pattern of control and access to fuel stocks is such that they will in fact be spending more in absolute terms, than wealthier groups, to satisfy their fuel needs.

We need, when considering possible solutions to this problem, to keep this specific social structural background in mind. At the broadest level, it might be thought (as suggested by Briscoe (1979) that the creation and distribution of greater amounts of tree fuel could exercise some positive effects on all groups. Directly, a lowering of the price of fuelwood through additions to supply should favour lower-income groups who tend to be large purchasers of this commodity. More generally, the availability of greater amounts of fuel in the aggregate should benefit the poor: as noted earlier, much of their fuel (residues, twigs, leaves, dung and so on) was gathered from the lands of richer farmers, but this traditional right of forage has been reduced of late due to overall pressure on fuel. A general reduction of pressure on resources should therefore benefit poorer groups.

But such value-neutral reasoning needs to be kept in some perspective. In the first place, the apparent size of the deficit between projected consumption and supply suggests that it will be some considerable time before feasible additions to supply could exercise much of a downward

price effect. Secondly, experience in Bangladesh shows that schemes direct
at market supply in general have a tendency to deliver maximum benefits to
richer groups, and minimal or zero benefits to poorer people. The dilemma
therefore, becomes one of devising schemes which will deliver at least some
significant measure of improvement in the fuel-energy flow to those most in
need within the social and political constraints that apply.

One set of options that might achieve some improvements in this
regard are fuel efficiency measures. There are now available designs for
much more efficient chulahs, or cookers, than are currently in use. There
exist practical designs for solar cookers and dryers which can be manufactur
from cheap, locally available materials. The principal disadvantage of thi
option would be the difficulty of reaching the poorer groups in society with
information. Its advantage is that it is principally a transfer of
information. There are no physical inputs to be siphoned off in various
counter-productive ways, nor expensive items to be purchased to operate the
system. This, to some extent, would overcome the probability noted by
Briscoe (1979) that distribution of premade devices would benefit mainly the
rich. If the principal of manufacture from locally available materials is
followed and only instructions for manufacture and use are distributed, ther
is at least a possibility that village elites will have less incentive to
interrupt the process of distribution.

A second set of options, which we will consider in the following
section of this Chapter, involves the use of Government lands, or follow area
for fuelwood plantation.

It is virtually an axiom in Bangladesh, and countries like it, that
when pressure is put on the basic necessities of life, the very large group
landless and poor rural dwellers will bear a disproportionately large share
the effects. Moreover, as the availability of organic residues contracts,
soil fertility will decline and the result of this can only be a further flc
of people into lower income and landless categories. We have accepted, in
this book, the extreme difficulty that any government of a country such as

Bangladesh would have in trying to enact significant, direct measures for land reform or income distribution. On the other hand, we have rejected the idea that a unified conspiracy in government exists, to keep present inequities in place. There is no strong reason, therefore, why the government of Bangladesh could not accept schemes aimed at fuel efficiency improvements, and involvement of poor and landless groups directly in fuelwood afforestation on government lands.

4.3 Forests and Rural Development

We are now in a position to discuss some of the more recent thinking and field projects, in the general area of forestry as a means of promoting rural development. There is now little doubt that, for the poorest group of countries at least, rural development which increases output, productivity and welfare of the agricultural sector must be a primary goal. Income redistribution and the generation of sustained increases in effective demands in LDC's will ultimately depend upon significant growth in their rural sectors. Forestry can assist this process: by providing much needed increases in a traditional fuel, (which adds directly to welfare, and also indirectly by relieving pressure on other combustibles with important alternative agricultural uses); by providing important opportunities (usually in association with agriculture) for settlement of shifting cultivators, and consequent control over erosion of land; by adding to (usually scarce) stocks of basic structural raw material; and by providing a range of additional income opportunities to rural dwellers.

But forestry can also impede the process of rural development, and this is an aspect which does not receive a great deal of attention in the literature. It can keep areas of potentially high productivity under low value-output usage in cases where officials of a forest service act to protect "their" forests from encroachers, to the exclusion of all other options. It can absorb scarce capital in plantation and other types of long term activity which are indefensible when appropriate LDC opportunity costs are imputed for that capital - particularly when no accompanying measures to provide short term, locally utilizable output from the resource are enacted(3). Management

of the forest resource principally for use by large scale industries can, by
process of cyclical reasoning be used to justify the continued operation of
those industries even when they are uneconomic - thus absorbing more valuable
capital and other resources. It is important to bear in mind these negative
possibilities, whenever considering forestry options.

It is also important to recognise that the mere orientation of
forestry activities towards rural development ends is no guarantee that all -
or indeed any - of the general benefits listed above for it will eventuate.
Problems of inapropriateness in the techniques implemented; of inadequacies c
the management and administrative structure; and, importantly, of unaccounted
for socio-cultural phenomena cutting across the projected benefit stream from
the project, are all very real possibilities.

With these reservations in mind, we will now review some major
forestry or agroforestry projects which have been attempted in various places
in the world.

4.3.1 The Philippines: a small farmer tree-growing project

In 1957, the public forests products corporation of the Philippines
(Paper Industries Corporation of the Philippines (PICOP)) embarked on a
programme to produce pulp and paper at a site on the northern side of
Mindanao. Although the Government attempted to ensure tenure of the forest
supply area for this operation, it became apparent that this was not going
to prevent encroachment onto the forest area for cropping or grazing
purposes, by local rural dwellers and also by shifting cultivators living
within the forest.

The shifting cultivator problem has been handled by offering the
cultivators three alternatives: employment within the corporation; a cash
settlement on the understanding that they will move elsewhere; resettlement
in places set aside for permanent agriculture, with building materials and 2
hectares of land being the basis. According to Arnold (1979), by 1977
one half of the identified shifting cultivation families has been
resettled.

The encorachment problem from small farmers was handled rather differently: the objectives were not only to prevent the encorachment, but to ensure that the small farmers involved could earn sufficient to reduce social tensions between them and corporation employees brought about by income differentials. The basis of the plan was the encouragement of farmers to grow trees as a cash crop, for sale as pulpwood to the corporation. Thus, credit is upplied through the Development Bank of the Philippines (supported by a World Bank loan). Farmers who do not have title to the land they occupy are assisted in acquiring it, and extension and advice services are supplied. Farmers are encouraged to devote about 80% of their land to trees - mainly (in this area) <u>Albizzia falcataria</u> - a fast growing species particularly suited as a pulp furnish for newsprint.

According to Arnold, this scheme has been successful: some 16 609 hectares of trees had been planted by farmers by 1978, and the economic rate of return to the project is in the order of 25 per cent per annum.

A second phase of the project, broadening it to small farmer tree planting activities throughout the Philippines is now in progress. According to the World Bank, a significant proportion of these will be based on fuelwood and charcoal plantation, and leaf-meal (from the Ipil-ipil tree).

There are some limitations to these sorts of activity. They rely to a large extent on the existence of an effective market: in the case of the fuelwood and charcoal outputs, the intention is to utilise these in power generation plants. So, some of the problems with industrial development in LDC's which we have examined in this book will apply: the pulp and paper industry in the Philippines does apparently show signs of this problem, and, because of the reliance on an industrial development of some kind, areas outside the economic supply zone for these cannot be included in the scheme. For this reason, Spears (1980) warns against the adoption of this approach as a solution in all forest areas where shifting cultivation is a problem.

The project has two elements of particular interest to us here. First, it is based on a cash crop approach, and whatever the infrastructural disadvantages and constraints on this method of operation may be, it does offer the opportunity of quick, visible returns to those involved - an

important factor in generating the attitudinal changes necessary to success.
Second, the project (in its initial phase) is a casw where an <u>existing</u>
industrial complex adapted its method of obtaining the raw material resource,
in such a manner as to involve the local rural population. This approach
could have considerable application in those many parts of the world where
large forest processing industries are already in place.

4.3.2 <u>Northern Thailand</u>

Northern Thailand has a tribal shifting cultivator problem somewhat
similar to that of Bangladesh. In the highland areas of the region the
consequent destruction of forest cover has become a serious threat both to
the water supply to the main rice growing area on the plains, and to the
native teak resource of the forests. A further complication comes about
through the fact that the tribal population has come to rely on the
opium poppy as their main cash crop, and the bulk of the product enters
the illicit world drug trade.

A number of measures based on integrated rural development, crop
diversification and reforestation are currently being tried. In the
forestry case, the Forest Industry Organisation which is responsible for
the management of the teak resource has acknowledged that it is no longer
feasible to settle all forest dwellers outside the forest areas (even if
the people themselves could have been convinced to leave their traditional
areas). The approach, therefore, has been to settle people in areas in
the forests where agriculture can be practiced: settlement is encouraged
by provision of roads, schools, electricity, tap water and other facilities.
The people are also offered employment in reforestation and other forestry
work. Much of the reforestation work is done on the basis of the agro-
forestry system known as <u>taungya</u> where tree seedlings are interplanted with
agricultural crops until the trees fully occupy the sites. Farmers are
allowed the agricultural produce from the area, in return for establishing
and tending the trees.

Promising as this approach is, it is apparent that alone, it would not
solve the shifting cultivator problem, because where this form of land
use becomes a problem, the fallow period before land is cleared and burnt aga

is too short to allow trees to mature. This is why, in the Thai case, the opportunity to earn income from eployment by the forest authority has been added into the scheme. Another important element which has been learned through experience with this scheme since its inception in 1968 is that security of tenure over the agricultural village land must be offered the villagers, before they will co-operate in the programme.

4.3.3. Kenya

As described by Spears (1980), Kenya has utilised a rather similar system to the Thai taungya scheme to re-afforest native forest areas with fast growing pines and cypresses. Perhaps the most interesting feature of this programme is that it has apparently been successful on a fairly large scale: to date, some 160 000 hectares of plantation have been established, and although these amount to only 10 per cent of the previously naturally forested area, planatations now supply 80 per cent of Kenya's domestic and export wood demand. This is relevant to our general comments on the world wood scarcity hypothesis in Section 3.5, Chapter 3, above.

4.3.4. China

The development of the communal forests of the Peoples' Republic of China certainly represents the largest programme of its land in the world. In 1949, forested areas had been reduced to 8.6 per cent of the land area of the nation. Since that year, some 28 million hectares of forest have been planted - three quarters of it by the communal farms, or 'production brigades'. Little is known about the quality or stocking of the resource but, even allowing for deficiencies in these areas, the nominal increase in the total planted area of the nation to almost 13 per cent represents a considerable achievement. Some of the planting (see Hillis (1982)) has been used to considerable effect for soil stabilisation and improvement along the south east coast of the country, and on the highland areas of Inner Mangolia.

Although no published or systematic evidence on the matter of rights, tenure and so on is available, some informal observation of the communal forestry scene in China suggests that individuals involved are usually able to identify areas on trees which are 'theirs', and it may be that motivation for establishment and tending by rural dwellers is being secured in this way, within the communal system.

4.3.5 Bangladesh: a community forestry project

In sub-section 4.2.2. above, we discussed the impending fuelwood crisis in Bangladesh, and we remarked upon the pressure which is being placed upon the resource. As is described in detail in Douglas (1980), the intensity of the problem varies from place to place in Bangladesh - but even so, two general claims can safely be made on the basis of existing data:

- the homestead forests, taken as a whole, are not sufficiently productive to cater for present and predicted demands upon them;

- the national forest resources of the country presently only supply about 20 per cent of total volume used for all purposes - and, even at this level of production (combined with the encroachment and illicit fellings problems) seems to be in a state of decline.

Under these circumstances, there is little possibility (in the short to medium term, at least) of utilising existing national forest reserves to meet fuelwood and basic building material shortfalls much beyond the level that they already do so. Therefore, the only options are to create a larger resource in the rural area, or to tolerate further declines in per capita availability and consumption of wood. As we saw in our discussion of the consumption aspects of rural fuel in Bangladesh, the latter option would imply further declines in rural welfare, most likely impacting heavily on the low income group.

Regional consumption and supply figures (see Douglas (1981)), Hammermaster (1981)) show that the problem is worst in the northern and western areas of the countries: - these areas are drier than other regions, and have been more affected by recent droughts - one result of which has been a heavy rate of felling of trees. Based on these results the Asian Development Bank in partnership with the Bangladesh Forests Department was, at the time of writing, in the process of initiating a broad community forestry project for the north-western area of the country.

The project is relatively large - involving expenditure of some $US20 million - and it covers a region in which resides almost a third of the population of the country. In its original conception (this may alter with experience) the project has three major elements:

- provision of tree nursery and extension facilities throughout the region, with the purpose of enhancing homestead forest plantings;

- funding and equipment for plantation of roadsides, railsides, canal embankments and other public areas;

- some regular plantation activity in degraded forest areas still nominally under the control of the Forests Department on the plain areas of the country. Although a small experimental area based on the co-operation of poor and landless local dwellers is included in this scheme, for the most part it was to involve the Forest Department in a traditional government based plantation effort.

At first glance, a project of this nature might seem generally beneficial for Bangladesh:

- it is aimed at the right area of the country, so far as the shortage problem is concerned;

- it seeks to provide more wood within the rural environment, where it has important basic uses;

- it seeks to utilise empty (or apparently empty) areas for the purpose of plantation, rather than attempting to transfer much-needed agricultural land to that purpose;

- it promises certain environmental benefits, through re-planint of degraded and eroding previous forest sites.

However, in our view, the project (as presently designed) suffers from some problems which ultimately will threaten its whole success:

- by directing most effort at enhancement of homestead plantings, the project is automatically biasing the delivery of most benefits to larger landholders. To be fair, some anticipation of this has been made and an attempt to limit assistance to smaller farmers will be made. However, it seems most unlikely to us, based on precedents in Bangladesh and elsewhere, that larger, better organised landholders will be able to be excluded from a major share in benefits. Landless or very small landholders certainly will not benefit from this aspect of the scheme, for the simple reason that they will not have land to spare for the purpose. In fact, land availability on the homestead sites in general is probably far more limited than the project seems to assume. Spears (1980) has remarked upon the difficulty of finding room for tree planting in crowded countries, and on the reluctance of small farmers especially to do so. In our experience, the homestead sites of Bangladesh already seem very fully occupied: changes in species composition and in quality of the growing stock could be used to increase output, but these are __long__ term options;

- although the project will seek to use landless and poor rural dwellers as labour, to some extent, it will provide few other benefits to them. There appears to be little intention to involve such groups in management of the resource created - including a share in the proceeds from sale;

- most of the supposedly fallow areas of Bangladesh - roadsides, railsides and so on - are in fact heavily used, for small scale
  cropping or grazing. These uses are not totally incompatible
  with tree-growing so long as those who presently utilise such
  areas are contacted, and are offered some incentive to protect
  established trees;

- apart from some marginal plantings on degraded Forests Department
  areas, the project ignores conversion of the quite significant areas
  of remaining government forest on the plain land area to fast
  growing plantation. Presently, these sites are occupied by slow-
  growing Sal (<u>Shorea robusta</u>) forests, which (not surprisingly)
  are subject to considerable pressures of encroachment and illicit
  removal, and are the source of much tension between the Forest
  Department and rural dwellers as a result.

There is, in fact, little point in attempting a community oriented forestry project in Bangladesh, because there is no strong communal structure in rural society. We do not, we hasten to point out, recommend aboundonment of this project - nor even of those elements of it we have criticised. Rather, we would advocate a reversal of its priorities. It should aim at enhancement of income opportunities for poor and landless groups, rather than an aggregate objective to create more wood in the rural area. Thus, the involvement of poor and landless groups - on a financial share basis - in re-plantation of the government forest areas, should be a major activity, whilst the enhancement of homestead plantings should be a minor one. Planting of government land on this basis could, apart from anything else, provide quite a handsome return to government on the investment we attempt some nominal calculations on an agroforestry approach to this in Appendix C.

We are not arguing that a re-orientation of priorities along the lines we suggest will solve all problems. Any attempt to direct benefits at the poor in a country such as Bangladesh will inevitably encounter problems. Other groups in rural society can hardly be expected to refrain from attempts to expropriate some benefits for themselves: indeed, our suggestion that enhancement of homestead woodlots be retained in the project is to that landholders also perceive some benefit to themselves from the project activity. But at least the operation of a large part of the programme on government land would allow some measure of control and would not involve any specific transfers from richer to poorer groups.

In the final analysis, it would seem to us preferable to attempt projects which at least address the critical distribution problem, in preference to those which cannot make any impression upon it.

4.4. Is Agro-Forestry Always the Answer?

In countries where land is short and where there is also a requirement for afforestation, it clearly makes sense to utilise forestry sites to the maximum extent possible, and trees do not fully occupy sites until some years have elapsed. But this does not suggest that it is always logical to attempt agriculture and forestry on the same areas in perpetuity. Even in cases where forestry and agriculture are to be practiced by the same people, as part of re-settlement schemes, the two activities might occur in different areas: the option of employing people in pure forestry, and allocating separate areas to them for subsistence or even cash crop agriculture has had considerable success in some areas.

Spears (1980) points out some trend towards a specialisation into monoculture tree crops in some areas. Although he points out that the role of the mixed agriculture forest homestead site is not likely to alter in the humid tropics, he observes that where small farmers do begin to expand their cash crops, they are likely to do so in monoculture operations, rather than agro-forestry ones. He cites the increasing tendency for cocoa cultivation to be on a pure crop basis, and a tendency in Malaysia for tree fruit crops to follow the same course, as examples.

NOTES (PART 2)

"The ideas of economists and political philosophers, both when
they are right and when they are wrong, are more powerful than
is commonly understood. Indeed, the world is ruled by little
else. Practical men who believe themselves to be quite exempt
from any intellectual influences, are usually the slave of some
defunct economist. Madmen in authority who hear voices in the
air, are distilling their frenzy from some academic scribbler
of a few years back. I am sure that the power of vested interests
is vastly exaggerated when compared with the gradual encorachment
of ideas."

The bulk of data referred to in this section comes from a
forthcoming paper by S. Parsons of the Australian Bureau of
Agricultural Economics.

An interesting case in point is the teak plantings which have been
(and to some extent remain) the mainstay of official Government
plantation activity in the hill areas of Bangladesh. Even when
this is done on optimal sites (and many of the plantings there
are not), it is difficult to see how Bangladesh could justify
continuing to invest in projects in which the final crop will not
mature for 60 years or more. Particularly when, according to at
least some writers (see, for example, Keogh (1979)), the market
prospects for teak are not universally favourable. Foresters in
Bangladesh sometimes argue that the reason the teak plantations
in the hills are now subject to the same problems of encroachment
and illicit (premature) felling by local dwellers as the native
forest areas, is that the people do not understand the significance
of the resource. We are bound to observe that even if the logic
were explained to them, they could be forgiven for not being
particularly impressed by it.

PART 3

CONSTRAINTS AND CONCLUSIONS

"It's true our Project is aimed at the Best;
There'll be more for them, less for the rest.
But it's not our problem, please don't raise it:
We're only here to pre-appraise it".

Additional verse to <u>The Development Set</u> (anon),
by J.J. Douglas.

Chapter 5

NEW APPROACHES TO DEVELOPMENT

5.1 Some Political and Historical Aspects

There are a number of feasible explanations for the rather disappointing progress in the developing world since the Second World War. We cannot, in this book, hope to resolve the considerable areas of controversy and conflicting indications that still characterise the theory. Rather, we have been principally concerned with examining some of the more commonly argued remedies for underdevelopment. We have seen that these have evolved from an emphasis on accelerated industrialisation as a means of transforming backward and unproductive economies; through a period of heavy emphasis on aggregate improvements in agricultural output; to the present point where the mainstream of post-modernisation theorists and neo-marxists have drawn rather closer together on the importance of income distribution in the sustainable growth process.

At the same time, there has arisen in some quarters a questioning of the whole rationale for development assistance as a means of alleviating poverty. On the basis of our consideration of economic growth and income distribution in Chapter 2, we would be justified in suggesting that aid <u>as it has been applied</u>, has achieved relatively little in this regard. The ways in which it has been applied, the scale of it (relative to the size of the problem) and the distribution of it <u>between</u> recipient countries, have mitigated against its effectiveness in relieving extreme poverty. In the poorer of LDC's, the situation is still one of increasing rather than decreasing inequalities in welfare, status and power. The rural sectors of these countries are still receiving an inadequate share of the development capital stock, whilst the migration to the cities that has been one consequence of this has led to urban unemployment and congestion becoming major problems for many of them.

Johnson (1975), in an assessment of development, argues that the development problem was in fact misconceived in two important ways. First, it was defined in terms too narrow to include relevant social and political aspects. Second, it placed reliance on the manipulation of a few strategic economic variables which were assumed to possess leverage over the rest of the system.

Many of the economists who become involved in development were conditioned by events of the Depression to regard the market for primary products as inherently unstable, and to consider the capitalist economic model a failure, in terms of organising production and consumption, and maintaining full employment and growth. The answer to this unreliable system was seen to be <u>planning</u>. Early experiences with planning in Europe had involved emphasis on expansion of heavy industry as a engine of growth - and much success with this approach was in fact obtained under the Marshall Plan for reconstruction of Europe. With hindsight, it is relatively easy to see why a war-torn Europe was an extremely poor model for underdeveloped economies: the industrial and commercial organisation and the skilled manpower for industry were all largely intact in Europe after the War, even if some capital had been destroyed. By replacing the capital - which in effect was the objective of the Marshall Plan - it was certain that the rest of the development process would follow. But in LDC's, it is not, as we have seen, merely capital which is lacking - it is <u>all</u> of the elements necessary to produce high economic growth and productivity, along the lines of Western economies.

A second important oversight in earlier development thinking (and it is one which persists in the development community even today) was a political one. It was assumed, in effect, that LDC's had gone through the process of political integration that had occurred in Western nations; that the nations involved were thus capable of transformation of society by peaceful means. In this respect, Johnson draws an important distinction between what we might term 'imperial' territories, and colonised ones. The former type, ruled by a thin upper crust of foreign civil servants and selected, educated nationals, have been characterised since their independence from this system by instability, often to the extent of civil war. This has been brought about because, once the threat of <u>external</u> force (the principal means of government under the imperialist phase) was removed, local disparate groups were unable to agree on major national objectives or even, in many cases, that they <u>were</u> a single nation. Moreover, the administrative systems of these countries tended to disintegrate, having been based on foreign administrators (or on the educated upper level of national civil servants who tended to be unpopular, and vulnerable when the colonial power was removed).

Colonised territories - that is, those settled by immigration from the imperial centre (or other European countries) - have a quite different history. In them, indigenous people previously settled in the areas were either killed off, or isolated into reservations. Whether because of a healthy climate for avarice, or through some sort of 'frontier' mentality, such areas have tended to become high growth areas (notwithstanding whatever social and cultural problems from the colonial hangover they might have).

Whilst this analysis certainly does help to explain political and administrative breakdown, the persistence of inequalities, and the failure of European-based development strategies in many countries, we are not so confident as Johnson appears to be that it is the complete answer. It does not explain, for example, why countries such as Thailand (which was never colonised) has today many of the typical LDC problems - nor why some countries where a pervasive imperialist presence was once a feature (such as Singapore) have managed to achieve high growth and redistribution. It can even be argued that India - the classic example of the large, overpopulated LDC - owes fewer of its problems to the Raj than may once have been thought; the influences (good or evil) of the Raj on the general structure of Indian life and culture may in fact have been fairly marginal. Certainly it seems to us rather unlikely that things would have been very different for the great bulk of rural dwellers, regardless of who ruled in Delhi and the district offices. Nor, for that matter, might the fate of the Maharajahs have been very different.

In any event, it is apparent that in societies where social divisions are deep, it is highly likely that processes such as massive industrialisation will lead either to an amassing of wealth by an urban based power elite, or by foreign owners. Ultimately, this may give way to more balanced distribution, as we saw in our discussion of Dependency Theory. This will depend on the rate of political maturation in the country itself. For democracy to grow and survive, thre must be a concensus about what the purposes of the nation are, so that changes of political parties involve only marginal shifts in policy. Democracy will not persist if it is used as a disguise for control by a small elitist group - nor if it means severe economic and social losses for the losing party.

5.2 A Perspective on the Economics of Development

The major point which emerges from the above discussion is that much of what has gone on in development theory and practice in the past has been based upon postulations on political maturity and government intent which have not proved to be sustainable. Our view of the matter, as we have suggested in our analysis of the Second Five Year Plan for Bangladesh in Chapter 1, is that it is not helpful to assume either that the government of an LDC is a unified and highly organised conspiracy to preserve elitist privileges at all costs, nor that it is a bland administrative machine fully prepared to cooperate in a whole range of development projects. We will return briefly to this issue in Chapter 6.

One conclusion which is commonly drawn, in the face of the failure of many developing projects and the role of governments in these, is that the role of LDC governments should be minimised: this is the antithesis of the earlier emphasis on planning as a means of solving problems:

> " I believe that the mass of the evidence on economic development accumulated in the past twenty years shows that people do respond sensitively to economic incentives and that market signals are generally better guides to economic opportunity than governmental judgements. One implication is that government should be careful about intervening extensively in the competitive process. "
>
> Johnson (1975)

However, in the situation where a laissez-faire approach would imply a continuation of trends towards impoverishment of increasing numbers of people, we must question this attitude. In highly dualistic unequal societies, if governments do not do something to create the economic opportunities and incentives to which Johnson refers, then who will? In any event, the reality of the development process (or the lack of it) in many LDC's, is that it depends on a symbiosis between large development assistance agencies, and governments. Neither are likely to abdicate their positions, and thus in practical terms a re-direction of their activities would seem to hold most promise of real change. And, obviously, perceptions in these quarters <u>are</u> changing, even if manifestations of this in the real life situations facing most people in LDCs are presently few. Most of the observations we have drawn in this book are from this viewpoint that change must come from existing structures. This is not complacency, for there is little in the developing world to be complacent about. Rather, it is based on a general belief that most successful revolutions - of whatever type - are preceded by an evolutionary phase which involves a process of altering perceptions, improving knowledge and information, and raising levels of mass awareness about the real nature of the problems present.

To pursue the aim of sustainable growth, rather than one of industrialisation for its own sake, LDC policy makers will need to reconsider very carefully the merits of the disguised unemployment argument as a defence for setting up import replacement industries behind a network of tariff and non-tariff barriers. The most acceptable view, on current evidence, would be that deliberate industrialisation via import substitution is probably a largely inappropriate strategy for most developing countries. Hughes (1971) observed that South East Asian countries which were concentrating in the late 1960's on import replacement (the Philippines, Malaysia, Thailand) had experienced slow-downs in their industrial growth rates. Whilst the trading situation faced by developing countries is by no means ideal, neither are the disadvantages and disincentives insurmountable. An obvious result of these observations should be that countries which have pursued an <u>export</u> strategy should have performed relatively better. Using the examples of Singapore, Taiwan and South Korea, Bhagwati and Kreuger (1973) have in fact noted that this is the case. They conclude that, although economic intervention in countries pursuing the export option has been no less chaotic, the economic costs have been less than in import replacement countries.

The basic factor, when considering the industrialisation objective in a developing country, is the relationship between agriculture and industry. Within the constraints of the export strategy there will always be some need for a developing country to manufacture items for domestic consumption. It is likely that industries which supplement agricultural production, rather than attempt to substitute for it, will meet with greater success, and in this context we need to bear in mind the observation that appropriate technology refers not only to the means of production, but also its purpose in overall economic terms.

It should by now be clear that there is nothing inherently inferior about the agriculture sector, and this view, whilst by no means yet general in LDC's is now at least being argued there: "The traditional concept of development through the growth of the modern sector by withdrawing labour from the agriculture sector is no longer accepted as a guide for development for a surplus population country such as ours" (Rahim (1978)). It is certainly true that impressive output growth rates can be obtained from investment in manufacturing, starting from a low base. But the costs of this, in terms of agricultural productivity foregone, have often been inadequately measured, or not measured at all. We have argued in this book that industry which does not have some direct positive linkage to the agriculture sector, or some considerable potential to efficiently generate export income (and this in an environment where some of the resultant gains can be effectively re-distributed to the agriculture sector) should at the least be regarded with a good deal of suspicion. The concentration of a large proportion of a nation's resources in agriculture is not at itself a cause of poverty: low growth and productivity in the sector is the cause. Lipton (1980), in analysing the well-known problem of rural migration in less developed countries, emphasises the need for greater attention to the agriculture sector: "....It helps little to provide more urban jobs, to which more than proportionately increasing number of villagers will flow, increasing the dearth of rural leadership and skills without improving the urban situation. Rationalisation of migrant flows require more rational policies towards agriculture."

So far as the distribution of income within the agriculture sector is concerned, we have suggested that whilst there is not yet any absolute proof that the marginal productivity of capital in the rural poor sector will be high, nor is there convincing evidence for the necessity to sustain, or exacerbate the destitution of the rural poor in order to obtain high levels of economic growth and development. In any event, given the levels of hardship currently being manifested, it is reasonable to argue that, quite apart from the moral and humanitarian issues involved, it will soon become impossible for many LDC's to persist with policies, which, in effect, make the rich richer and the poor poorer. Political disruptions of economic life or, to offset these, crippling social welfare payments, are the obvious consequences. In LDC's human capital is the one type which is frequently available in abundance. Rather than being run down, it should to the maximum extent possible be substituted for scarcer inputs.

Meier (op. cit.) cites a number of studies to support the argument that when the problems of rural productivity and prosperity are attended to, the non-rural sector of the population will <u>as a result</u> become larger and more prosperous. It is reasonable to suggest that initially agrarian countries which have developed rapidly in the (relatively) recent past – and the examples of Japan, South Korea and Taiwan are relevant – have initially achieved a rapid expansion in agricultural output, largely via investment in land-saving techniques, before the industrialisation process commenced (or at least in its very early phases).

Economic development should not be regarded as synonymous with economic independence, industrialisation, import-saving, self-sufficiency or any of the similar catch-words that have been frequently applied. In very poor countries, the achievement of an agricultural surplus and reduction of the level of absolute destitution must be the primary objectives of development. Industrialisation should be pursued to the extent where it materially assists this process, either directly through the generation of inputs and incentives, or indirectly through the <u>efficient</u> generation of export income for the country as a whole, and only then if some effective means of distribution of gains to the rural sector exist. Otherwise, the possibility of enforced industrialisation bringing about reductions in food output is very real.

Chapter 6

FORESTRY AND DEVELOPMENT: THE PROBLEM RE-STATED

It has been a theme of this book that forestry is no more exempt from the implications of a general theory of development than any other sector of an LDC. Indeed, as a land based sector, with variable elements of complementarity and conflict with rural production, there are good reasons for examining its performance and modus operandi even more closely than some other sectors.

The result of carrying out such an examination, in many LDCs, will be to reveal a sector which exemplifies the problems of dualistic development strategies. At the raw material level, official government sector administrators will very often see the basic task of government forestry to be provision of raw materials to large scale industries, and in the longer term to establish areas of high quality timber plantations, often in pursuit of a projected export market. The official forestry sector in the country will commonly be short of personnel and equipment. Yet the modus operandi of restricting entry and usage of national forests (or attempting to do so) and operating purely on the basis of hired departmental staff places greatest pressure on the adminisatration. The result has been to create an economic vacuum in the forests, into which the surrounding population naturally flows. The task of keeping them out adds further to administrative burdens, and the standard of management - particularly the ability to experiment and innovate - drops even further. Thus, the national forest lands end up simply not pulling their economic weight. Such output as they do produce is of little or no relevance to the surrounding rural population who, in consequence, have no interest in co-operating with the official forestry authorities in managing the resource to maximise production.

The logical requirement of the forestry sector, in such countries, is usually rather different. Westoby's statement of a forestry desideratum (quoted on the facing page to Part 2 of this book) is a more defensible approach. However, it is necessary for us to attempt some consideration of what the implications of it might be in LDCs.

The effects might not be universally positive: we should not lose sight of the fact that in many LDCs, the social structure is rigid and highly stratified. The literature on community forestry frequently misses this point, or submerges it in blandness. There is, in fact, a deal of dubious presupposition in the very term 'community forestry'. Particularly in the situation where there is a dynamic shift in the economic status of many people - as there is in many LDCs with their pattern of burgeoning landless and destitute populations - the divisions between income groups are deep, and hostile. Community forestry will be just as vulnerable to disruption or paralysis from this factor as have been many other schemes aimed at rural community development. If, for example, we design a project which directly involves poor and landless people in fuelwood production (as suggested for Bangladesh in Chapter 4 above), then it is relatively simple for outsiders to the rural sector to see this as beneficial to everyone: the poor by direct additions to their income; farmers in general through improved demand (from the rural poor) for their products and through reduced pressure on their agricultural residues for fuel (thus allowing its alternative use for fertiliser, mulch and fodder); the country as a whole through enhanced political and social stability, and through the ultimate effect of improved, stable incomes on population growth, and so on. But, one immediate effect of increased fuelwood supplies in the rural areas might be a reduction in its market price. Existing <u>sellers</u> of fuelwood - larger land holders with surplus trees on their property, for example - can be expected to disapprove and to use their influence accordingly. In some countries, fuelwood may be the resource keeping some <u>smaller</u> farmers viable, and the price effect of additions to fuelwood supplies might force them into the landless category. It would certainly not be the first time that schemes conceived to assist one group in an LDC have had the (often unexpected) effect of worsening the plight of another.

Another problem could arise in the form of resentment in the unassisted group. In some societies (perhaps most, when all is said and done) there is a subliminal constraint on re-distribution: we would certainly not be the first to observe that it is part of human nature to instinctively resent the betterment of others, particularly when the comparison with oneself is fairly close (as it very often is in rural society in LDCs). This is no reason, of course, to abandon hope of achieving progress nor to indulge in excessive philosophical agonising over what to do about this aspect of human nature. However, it does emphasise that whenever intervention into the status quo is proposed, the initial sociological investigation and education processes need to be applied as much to these who will not benefit (immediately) as to those who will.(1)

What will be the effect of the policy on existing forest industry? Many LDCs have quite large forest products industries. It is all very well to suggest, as Westoby does, that the export option for LDCs must be looked upon with some suspicion, or at least caution, and we have discussed this point in general in Chapter 1. But what of the larger industries that exist now? Many of them are export-oriented, for the reason that domestic demand is insufficient to support them, whilst to scale them down to smaller operations would invite, or exacerbate the well known scale economy problems that affect most forest products industries. It is more likely, therefore, that LDC"s with significant forest processing capacity in the modern sector face a much harder choice: to persist with production, with all the attendent profitability problems and possibly resource problems as well; or to close the industries down, with the resultant social and political difficulties involved in disemploying large numbers of industrial workers. It seems probable to us that the necessary structural changes in LDCs to bring the rural forest development strategy into play will necessitate a phasing out of at least the most unprofitable of the large scale industries.(2) But we should be under no illusion that this will be at all easy: industry closures never are, and the welfare problems they create in poor countries, where both unemployment benefits and alternative employment opportunities are limited or non-existent, are severe. It is of little comfort to an LDC government, no matter how committed it may be to economic reform, to contemplate the advantages of employing perhaps three people in the rural area at the cost of one job in the industries, when that job happens to be held by a member of the vociferous and politically active urban minority.

What are the administrative constraints that apply? Obviously, there will be many, and they will vary in type and intensity from one country to another. All we can attempt to do at this stage is to question some of the common assumptions made about LDC governments, administrations, and policy change.

In Westoby's (1978) view of forestry development (and it is one commonly encountered in general economic literature as far back (at least) as Pareto (1935)) many LDC governments exist simply as instruments of exploitation of the rural masses. This implies a definite will, and capacity on the part of the LDC government and administration <u>as a whole</u> to put such tendencies into effect. On the other hand, the FAO, in its suggested strategy for the forestry sector, simply assumes a predisposition on the part of the LDC government <u>as a whole</u> to carry out the necessary measures to transform forestry.

Both assumptions - the one seemingly cynical, the other mechanically optimistic - are in our view far too general to be of use, and in any event are naive. Both fail to ask the vital question: what <u>is</u> the government, in the LDC context? Is it the political group in power at the moment, with its upper level advisors? Is it the (usually) longer established, and more traditional bureaucratic elite, operating at the sector administration level? Or is it the field operatives, who - especially in a sector such as forestry - form the connecting links between the government and the people? The point we are leading to, of course (and we have already spent some time on it in our Bangladesh case study) is that very often in LDCs, different levels of the government operate on very different assumptions about what needs to be done in the economy. The elected officials and their senior advisors <u>may</u> very well understand, and <u>genuinely</u> believe in current arguments in support of income distribution reform, the priority of rural development and so on. But this is no guarantee that the sector level administrators agree. Long established patterns of sector administration may be so ingrained as to lead bureaucrats to <u>perceive</u> what they are doing as being for the best, and as being consistent with government policy, even when an independent assessment would emphatically show this to be untrue.

In other words, in the situation where perverse and counterproductive expenditure patterns and policies in a given LDC sector are persisting, there is no automatic need to explain this solely on the basis of crippling corruption(3) and incompetence in the system: a lack of

conviction or understanding on the part of some decision makers is not
the same as either of these. Nor is there much point, we suggest, in
believing that convincing the planning level of government of the right
approach, and then providing the technical inputs needed for
implementation will somehow make the problem disappear. This belief,
often couched in terms of overcoming 'absorptive capacity' or
'implementation' constraints, is very often the basis of the technical
advice and inputs that constitute much of the aid effort of some of the
large development assistance agencies. But if sector administrators do
not believe in the edicts of their government (or do not interpret them
correctly) and if they are left to implement their own policies, then no
amount of technical assistance will change their minds. Indeed, it may
make things worse, by assisting the administrators to run even more
enthusiastically and energetically in the wrong direction.

Of course it may be that the problems of concept and communication
are at a different level - in some LDCs, the bureaucrats may be right,
and the politicians wrong. Or all may be misguided (although this, we
suggest, is probably much rarer than many development theorists seem to
think). Essentially, our argument is that the system of policy design,
decision and implementation for each country (and even perhaps for
individual bureaucratic groupings within countries) needs to be closely
examined before development strategies and projects are put into the
field (or, for that matter, into official policy). To assume the nature
of LDC governments to be monolithic can rarely, if ever, be valid.

What we have come to, in this general review of the role of forestry
in development, is a plea for an upgrading of the study of the political
sociology of the forestry sector in LDCs. Forestry has, as we have seen,
a direct and fundamental relationship to rural people, because it
occupies land, and because it is capable of providing products which, in
many countries, have a vital role to play in the energy/nutrition cycle
of the rural dweller. The changes required to systematically and
rationally transfer a greater share of the forestry asset into the hands
of those who need it most, and thereby to enhance the process of economic
development overall, are at base conceptual. They have less to do with
the techniques of improving productivity under current large scale
management practices, or even of administrative and institutional ways

and means of planting more trees in the villages (which seems to be all that many people now think is required), than with the character of the administration of the forestry sector itself in many LDCs.

We have argued above that it is too simplistic (usually) to assume a unified 'government' intent, in the practice of forestry in LDCs. What have evolved in many LDCs - particularly those with a fairly strong colonial heritage - are highly conservative and traditional forest services, which have no mechanisms for meaningful reference to the broader economic and political perspective of society. As Westoby (1978) puts it:

> 'For centuries much of the work of foresters went into creating and protecting royal and princely estates, extinguishing every kind of common right in the forest and enforcing exclusive property. The folk lore and oral tradition of many countries still makes the forester the people's enemy, the gendarme of the landed proprietor'

In many LDCs today, the princely estate or landed proprietor may have given way to the official forest service, but the effect remains the same. Originally, in these countries, the establishment of a conservation - oriented, centralised forest service might have been an acceptable way of introducing rational utilisation of a large, untapped resource. Perhaps for some sparsely populated and well forested LDCs, it is still a good system. For the heavily populated, resource-short remainder of LDCs, however, the persistence of a management ethos which systematically excludes or minimises the application of the only capital resource they have in abundance - human capital - to mobilisation of the resource, can only be counterproductive.

Had traditional forestry attitudes been formed entirely within the countries concerned themselves, then there may have been some possibility of their evolving with political change. Very often, however, the philosophical and academic bases for forestry have been acquired elsewhere - in Western of or Western-oriented institutions - and are inappropriate for LDCs. Also, much of the input, advice and exhortation which has come in the past from the large international agencies has tended to re-inforce these attitudes. Much of the technical assistance still being offered, in pursuit of 'institution building', 'strengthening the administrative framework' 'enhancing training and research capacity' and so on is in danger of achieving little more than perpetuation of this system, strengthening it technically against pressure for changes from below, or above.

So: the essential changes required are conceptual rather than technical. Even if this is accepted, the problem of transforming forestry will remain a sensitive, and in many ways an internal matter. If advice from the agencies is sought, then it should be provided. Most development assistance agencies persist with the myth that this is all they ever do - provide what is asked for. But the reality of the situation, as every operative in the field knows, is rather different: requests may officially come from governments, but their origins are often deep within the assistance agencies themselves. The problem seems to be the motives behind such offers: at the risk of over-generalising, it does seem that the greatest pressure on those in executive positions in headquarters offices of assistance agencies is to initiate projects - rather than to analyse; to consider priorities; and to devote most time to improving the quality and relevance of assistance. The competition for funds, which leads directly to disbursement, is intense, even within the major agencies, let alone between them. The capacity to disburse funds seems to be a more important criterion than the record of effective changes brought about in recipient countries. It is occasionally suggested that the large development banks are less inclined to this approach, because, as banks, some measurable return an investments must be generated. However, in our experience this is not so: a good deal of funding from the banks, in very poor countries, is on highly concessional terms amounting to little more than grants. Performance in the banks also seems to be judged primarily on the basis of project initiation. Even where a commercial loan is in operation, the fact that the recipient government meets its repayment commitments is no indication that the purpose of the loan has been met.

One effect of an emphasis on disbursement is that assistance agencies remain committed to their traditional area of expertise and technical speciality, because less effort can be devoted to changing these to suit the times. This, in our view, is a large part of the reason for the apparent dichotomy between the expressed aims and philosophies of aid (statements of which are easy enough to write) and the actual projects and programs in operation. Technical assistance, project oriented rather than concept oriented, continues to be given, whether or not the actual problems in the LDC recipient are amenable to this approach. The achievement, as we have noted, is little more than to assist a given sector to pursue inappropriate policies.

Thus, the 'foreign expert' arriving on the LDC scene with his supply of techniques and his mechanistic terms of reference, may very quickly find himself pondering the problem not of <u>how</u>, but of <u>whether</u> to apply them. If he encounters problems similar to some of those we have described in this book, and if he interprets them as we have, he may be surprised to find himself out of phase not only with local sector administrators, but also with those in his own organisation. He may even find himself in the peculiar situation of acting as a buffer, rather than a conduit between his own organisation, with its acute appetite for project initiation, and the local sector decision makers who, if they are to make appreciable progress in re-orienting themselves and the sector towards more progressive policies, will need to do so in an atmosphere of relative calm, with access to sympathetic, unbiased advice if they wish to use it.

One of our objectives in writing this book has been to make the task of the so-called 'expert', deciding what he <u>should</u> and <u>can</u> do, a little easier. If, in the process, we have indirectly made the tasks of others working the field of forestry sector development a little more difficult, then so be it.

NOTES (Part 3)

(1) For the record, we would point out that the first problem noted here – that of market and price disruption – probably would not occur under the sort of conditions prevailing in Bangladesh. The second problem – that of the creation of resentment – certainly would apply given the subordinate dependent state of the rural poor at present. In fact, it was for this reason that we suggested, in our discussion of the ADB communtiy forestry project, that <u>some</u> elements of assistance to landholders for homestead forests be retained, and that the major thrust of involvement of landless groups should be on government forested land, rather than supposedly fallow lands which in fact are heavily used.

(2) In some cases, the option may exist to substitute more efficient and appropriate industries for these – thus solving the unemployment problem. But, unless a feasible export market for the new product exists, some care is needed when considering this alternative, lest it lead simply to a substitution of an official processing sector for the small scale unofficial one which may already exist to serve the domestic market.

(3) The question of corruption is a vexed one in the development literature. Our view of it is perhaps more sanguine than most: it was widely present in the early developmental stages of most Western nations, which developed nonetheless. Rather than considering it an absolute impediment to development, we are more inclined to the view that it is as possible to be corrupt while doing mostly the right things for the country, as it is to be honest while doing mostly the wrong things. In our view, it is impossible to eliminate corruption in government, except under rather rare political and economic conditions. It is possible, however, to control it, and to direct it towards progressive ends. Very few individuals are incorruptible, but equally, very few are so corrupt as not to wish to do some good, and be remembered for it.

References

Ahluwlia, M.S. (1974), "Income inequality: some dimensions of the problem" Finance and Development, September 1974.

Ahmad, N. (1976), A New Economic Geography of Bangladesh, Vikas Publishing House, New Delhi.

Ahmed, S. (1980), "Thoughts on Rural Development", Bangladesh Times, 5/11/1980.

Amin, S. (1976), Unequal Development Harvester Press, Brighton.

Arnold, J.E.M. (Undated), Lessons of Experience in Planning Forestry Development, Mimeo, Harvard University.

_____ (1979), "New Approaches to Tropical Forestry: A Habitat for More than Just Trees" Ceres (FAO), September-October.

Bangladesh Bureau of Statistics (1979), Economic Indicators of Bangladesh Vol. VI.

_____ (1980), 1979 Statistical Yearbook of Bangladesh.

Bhagwati, J.N. and A.O. Kreuger (1973), "Exchange control, liberalization and economic development", American Economic Review, Papers and Proceedings, May 1973.

Bhattacharya, D. (1977), "Flow of resources between rich and poor countries", 48th ANZAAS Conference, Melbourne, 1977.

Bienefeld, M. (1980), "Dependency in the Eighties", IDS Bulletin, Vol 12, No. 1, Sussex University.

Brandt Commission (1980), North-South: A Programme for Survival, Pan Books, London.

Briscoe, J. (1979), "Energy use and social structure in a Bangladesh village", *Population and Development Review* 12:79.

Byron, R.N. (1979), "An economic assessment of the export potential of Australian forest products", *Industry Economics Monograph No. 20*, Australian Bureau of Agricultural Economics, Canberra.

Byron, R.N. (1981), *Future Consumption of Wood and Wood Products in Bangladesh*, Field Document No. 4, FAO/UNDP Project BGD/78/010.

Cardoso, F.H. (with E. Faletto), (1979), *Dependency and Development in Latin America*, University of California Press, Berkley.

Clark, C. (1974), "Optimising the supply and use of market timber through market prices", *Australian National Bank Monthly Summary*, October 1974.

Crosswell, M. (1981), "Growth, Poverty Alleviation and Foreign Assistance", *Development Digest*, Vol. XIX, No. 3.

Dell, S. (1979), "Basic Needs or Comprehensive Development: Should the UNDP Have a Development Strategy", *World Development*, Vol. 7.

Douglas, J.J. (1981), *Consumption and Supply of Wood and Bamboo in Bangladesh*, Field Document No. 2, FAO/UNDP Project BGD/78/010, FAO, Rome.

Douglas, J.J., Bond, G., Connell, P., Ramasamy, V., Buckley, C. and R. Treadwell (1976), *The Australian Softwood Products Industry*. Bureau of Agricultural Economics, Canberra.

Douglas, J.J., Rahman, K., and Aziz A. (1981), *The Industrial Forestry Sector of Bangladesh*, Field Document No. 3, FAO/UNDP Project BGD/78/010, FAO, Rome.

Douglas, J.J. (1982), "Traditional Fuel Usage and the Rural Poor in Bangladesh", *World Development*, Vol. 10, No. 8.

Earl, D.E. (1975), Forest Energy and Economic Development, Clarendon Press, Oxford.

Eberstadt, N. (1980), "Recent declines in fertility in less developed countries and what 'population planners' may learn from them", World Development, Vol. 8, No. 1.

Ellis, H.S. (1958), "Accelerated investment as a force in economic development", Quarterly Journal of Economics, November 1958.

Faaland, J. and J.R. Parkinson (1976), Bangladesh: The Test Case of Development, University Press Ltd., Dacca.

Food and Agriculture Organisation (1974), "An introduction to planning forestry development", FAO/SWE/TF 118, 1974.

_____ (1981), "FAO's Medium Term Objectives of Forestry and Programme of Work for 1982-83", Forestry Planning Newsletter, FAO, Rome.

_____ (1981), Map of the Fuelwood Situation in the Developing Countries, FAO, Rome.

_____ (1980), "Towards a forest strategy for development", COFO - 80./3. 1980.

Frank, A.G. (1978), Dependent Accumulation and Underdevelopment, Macmillan, London.

Fried, E. (1975), Untitled item in "Scanning our future", Report from the NGO Forum on the World Economic Order, Carnegie Endowment for International Peace, New York.

Fontaine, R. (1981), "What is Really Happening to Tropical Forests?" Ceres (FAO) July-August.

Gane, M. (1969), "Priorities in planning", Commonwealth Forestry Institute Paper No. 43, University of Oxford.

Gomez, G. (1978), "Case History of a South American Paper Mill", Unasylva, Vol. 30, No. 122.

Hayley, D. and J.J.G. Smith (1976), "Justification and sources of funding of forestry operations in developing countries" in Evaluation of the Contribution of Forestry to Economic Development, U.K. Forestry Commission Bulletin No. 56.

Hammermaster, E.T. (1981), Inventory Results, Field Document No. 5, FAO/UNDP Project BGD/78/020, FAO, Rome.

Henderson, P.D. (1980), "Survival, development and the Report of the Brandt Commission", World Economy, Vol. 3, No. 1.

Hicks, N.L. (1979), "Growth vs Basic Needs: Is There a Trade-off?" World Development, Vol. 7.

Hicks, N. and P. Streeten (1979), "Indicators of development: the search for a basic needs yardstick", World Development, Vol. 7.

Hillis, W.E. (1982), "Forest Products and People - Some Thoughts on USSR, China and Japan" Australian Forestry, 45(2).

Hossain, Mahbub (1974), "Farm size and productivity in Bangladesh agriculture: a case study of Phulpur farms", The Bangladesh Economic Review, Vol. No. 1.

Hrabovsky, J.P. (1980), "The Energy Octopus" Ceres (FAO), November-December.

Hughes, H. (1971), "The Manufacturing Industry Sector" in Southeast Asia's Economy in the 1970s Asian Development Bank, Longmann, London.

Islam, M.N. (1978), "Study of the problems and prospects of biogas technology as a mechanism for rural development: study in a pilot area of Bangladesh", Progress Report No. 1, Dept. of Chemical Engineering, Bangladesh University of Engineering and Technology, Dacca.

Islam, N. (1979), <u>Development Planning in Bangladesh</u>, University Press Ltd., Dacca.

Jabbar, M.A. (1979), "Relative productive efficiency of tenure classes in selected areas of Bangladesh", <u>The Bangladesh Development Studies</u>, Vol. V, No. 1.

Jenkins, R. (1979), "Europe and the developing world", <u>ODI Review</u>, No. 1 London.

Johnson, H.G., <u>On Economics and Society</u>, (esp. Ch.19) University of Chicago Press, Chicago and London.

_____ (1965), "Optimal Trade Intervention in the Presence of Domestic Distortions" in <u>Trade, Growth and the Balance of Payments</u>, (eds Cove, Johnson, Kenen) North Holland Publishing Company, Amsterdam.

Kirk, D. (1971), "A new demographic transition?" in <u>Rapid Population Growth</u>, Study Committee, National Academy of Sciences, Baltimore.

Keogh, R.M. (1979), "Does Teak have a future in tropical America?" <u>Unasylva</u>, Vol. 31, No. 126.

Leipziger D.M., and M.A. Lewis (1980), "Social indicators, growth and distribution", <u>World Development</u>, Vol. 8, No. 4, 1980.

Leslie, A.J., "Where Contradictory Theory and Practice Co-exist"

Leslie, A.J. and B. Kyrklund (1980), "Small-Scale Mills for Developing Countries", <u>Unasylva</u>, Vol. 32, No. 128.

Lewis, W.A. (1963), <u>The Theory of Economic Growth</u>, Allen and Unwin.

Lipton, M. (1980), "Migration from rural areas of poor countries: the impact on productivity and income distribution" <u>World Development</u>, Vol. 8, No. 1, 1980.

_____ (1977), *Why Poor People Stay Poor: Urban Bias in Developing Countries*, Temple Smith, London.

Little, I.M.D. (1973), *A Critique of Welfare Economics* (2nd Edition) Oxford University Press.

Little, I.M.D. and J.A. Mirlees (1974), *Project Appraisal and Planning for Developing Countries*, Heinemann, London.

McGrannahan, D., Richard, C., and Pizarro, E. (1981), "Development statistics and correlations: a comment on Hicks and Streeten" *World Development*, Vol. 9, No. 4.

McGregor, J.J. (1976), "The existing and potential roles of forestry in the economics of developing countries" in *Evaluation of the Contribution of Forestry to Economic Development*, U.K. Forestry Commission Bulletin No. 56.

Maloney, C., Ashraful Aziz K.M., Sarker, P.F. (1980), "Beliefs and fertility in Bangladesh", Mimeo report, Institute of Bangladesh Studies, Rajshahi University.

Marramma, V., Pera, A., and P. Puccinelli (1979), *Rapporto Economico Sulla Cina, Prezzi e Redditi*, Boringhieri, Torino.

Matsui, M. (1980), "Japan's forest resources", *Unasylva*, Vol. 32, No. 128.

Mauldin, W.P. and B. Berelson (1978), "Conditions of fertility decline in developing countries 1965-1975" in *Studies in Family Planning*. The Population Council, Vol. 9, No. 5.

Marsden, K. (1970), "Progressive Technologies for Developing Countries" *International Labour Review*, May.

Meier, G.M. (1976), *Leading Issues in Development*, (3rd Edition) New York: Oxford University Press.

Ministry of Agriculture and Forests (1979), <u>Bangladesh Country Review</u> (WCAARD Review) prepared for FAO World Conference on Agrarian Reform and Rural Development.

_____ (1979), <u>Costs and Returns Survey for Bangladesh 1978-1979 Crops</u>, Vol. III (T. Aman Paddy) and Vol. III (Jute), Agro-Economic Research Section.

Montreal Engineering Company Ltd; Snamprogetti Spa (1976), <u>Bangladesh Energy Study</u> Administered by Asian Development Bank, under UNDP Project BGD/73/038.

Myint, H. (1973), <u>The Economics of the Developing Countries</u>, (4th edition), Hutchinson University Library, 1973.

Nautiyal, J.C. (1967), "Possible contributions of timber production forestry to economic development", Ph.D. thesis, Faculty of Forestry, University of British Columbia.

Pareto, M. (1935), <u>The Mind and Society</u>, New York: Harcourt Bruce & Co.

Planning Commission (1980), <u>The Second Five Year Plan 1980-1985</u> (Draft). Government of the Peoples Republic of Bangladesh.

Rabbani, A.K.M. Ghulam (1965), <u>An Estimate of Long Term Timber Trends and Prospects in East Pakistan</u>, Planning Commission.

Rahim, A.M.A. (1978), "Leading issues in rural development" in <u>Current Issues of Bangladesh Economy</u>, (ed. Rahim) Bangladesh Books International Ltd.

Rahim, A.M.A. and M.S. Uddin (1978), "Demand for money in Bangladesh - a preliminary analysis" in <u>Current Issues of Bangladesh Economy</u> (ed. Rahim), Bangladesh Books International Ltd, Dacca.

Rosenstein - Rodan, P.M. (1943), "Problems of industrialisation of eastern and south-eastern Europe", <u>Economic Journal</u>, June-September, 1943.

Rostow, W.W. (1956), "The take-off into self-sustained growth", Economic Journal, March, 1956.

Schiffer, J. (1981), "The changing post-war pattern of development: the accumulated wisdom of Samir Amin", World Development, Vol. 9, No. 6.

Shepherd, K.R. (1980), "Energy from the forests: an exercise in community forestry for developing countries", Agricultural Information Development Bulletin, U.N. Economic and Social Commission for Asia and the Pacific, Vol. 2, No. 2.

Singer, H.W. (1975), The Strategy of International Development, Macmillan, London.

Sila-On, A. (1978), "The transfer of technology", Unasylva, Vol. 30, No. 122.

Slavicky, J.S. (1978), Forest Inventory and Aerial Photogrammetry, Field Document No. 5, FAO/UNDP Project BGFD/72/005.

Spears, J.S. (1980), "Can farming and forestry co-exist in the tropics?" Unasylva, Vol. 32, No. 128.

Streeten, P., "Industrialisation in a unified development strategy", World Development, January 1975.

United Nations (1974), World Economic Survey 1973: Part One, Population and Development, New York, 1974.

United Nations Industrial Development Organization (1972), Guidelines for Project Evaluation, New York: United Nations.

Vaitsos, C. (1979), "Bargaining and the distribution of returns in the purchase of technology by developing countries", IDS Bulletin, October.

Von Maydell, H.J. (1976), "Effective policies for stimulating investment in forestry industries in countries with tropical forests", summarised in Evaluation of the contribution of Forestry to Economic Development, U.K. Forestry Commission Bulletin No. 56.

Waterson, A. (1965), <u>Development Planning: Lessons of Experience</u>, John Hopkins Press.

Westoby, J.C. (1978), "Forest industries for socio-economic development" Guest speaker's address <u>8th World Forestry Congress</u>, Jakarta, October, 1978.

____ (1962), "Forest industries in the attack on underdevelopment" in <u>The State of Food and Agriculture</u> FAO, 1962.

World Bank (1980), <u>Bangladesh Current Economic Position and Short Term Outlook</u> (Restricted) 2870 - BD. South Asia Programs Department.

____ (1980), <u>Energy in the Developing Countries</u>, Mimeographed Report, Washington.

____ (1979), <u>Energy Options and Policy Issues in Developing Countries</u>, World Bank Staff Working Paper No. 350, Washington.

____ (1978), <u>Forestry Sector Policy Paper</u>.

Zaman M. Raquibuz (1973), "Share Cropping and economic efficiency in Bangladesh", <u>The Bangladesh Economic Review</u> Vol. 1, No. 2.

Appendix A

THE ECONOMY OF BANGLADESH

Bangladesh is a riverine country of some 144 000 square kilometres, bounded to the West and North by provinces of India, and to the South East by Burma. The predominating land form is a flat deltaic plain, which drains the massive Ganges, Brahmaputra and Meghna river systems into the Bay of Bengal. The eastern edge of the country is a tract of low hill country, and this zone is an important division ethnically and politically, as well as physically, in the country.

About 83 million people now live in the rural towns and villages of Bangladesh: its remaining 6 or 7 million live in urban settlements of more than 100 000 population. The density of population in the country is now 625 persons per square kilometre.

A. Brief Historical Perspective

The beginnings of agriculture in the part of the world now occupied by Bangladesh pre-date the Aryan migrations to the sub-continent. In the 7th Century, the Chinese travellor Hsuan Tsang was one of the early visitors to the area who commented upon the regular, intensive cultivation of land, and the abundance of grains, fruits and flowers. In addition to rice, there are very early references to the cultivation of sugar cane, mustard, fruits and cash crops of betel and cotton.

A large area of eastern India come under the rule of the Mughals in 1212 and, according to Ahmad (1977), the region prospered throughout the mediaeval era as part of the Bengal Sultanate ruled from Delhi. During this period, the area become well known for its handicrafts, as well as for agriculture.

After the battle of Plassey, British rule become established throughout the area, particularly from the time of permanent settlement under Cornwallis in 1790. In the 19th century, under the impetus of commercial developments emanating from British and other interests,

considerable specialisation into indigo, cotton and jute growing
occurred. The first two industries declined eventually in the face of
competition from European equivalents towards the end of the century. In
the course of these developments, many of the old handicraft skills
become lost, whilst later, many of those who had acquired the new ones
become redundant, and were forced back to the land for a living.

The British vested great powers in Hindu Zamindars (large
landholders) in the area, despite the large number of Muslims in the
population.

In these factors - the return of large numbers of industrial workers
to the land, and the divisions inherent in the Zamindar system - can be
seen the origins of many of the chronic economic and social problems
manifested today in Bangladesh.

In 1854, the large new province of Bengal was proclaimed: it
included Orissa and Bihar, and most of what is now Assam. In 1905, Lord
Curzon re-partitioned the province. In 1947, as is well known, the area
became East Bengal (later East Pakistan), a province of Pakistan, under
the Radcliffe award that accompanied the independence of the former
Indian Raj from Britain.

The union with West Pakistan was unfortunate both politically and
economically. Up to 1962-63, East Pakistan had a trade surplus, but in
large measure this was used to finance the deficit in West Pakistan.
Faaland and Parkinson (1976) estimate that some 15 000 to 30 000 million
Rupees were transferred from East to West Pakistan in this manner from
1947 to 1968-69.

In 1970, the leader of the East Pakistan based Awami League, Sheik
Mujibur Rahman, won an absolute majority of votes in the Pakistan
election. However, instead of taking over the Prime Ministerhsip, the
Sheik found himself in prison: factions and rifts throughout the East
broke out into a full scale and bloody revolution in 1971. This war,
although of short duration, was extremely costly in terms of lives and
property. Whether or not as a result of Indian intervention (there is
some contention on this subject), the West Pakistan forces were driven

out in 1971, and Sheik Mujibur Rahman took over as Prime Minister of the independent nation of Bangladesh. From this point on, the country suffered a number of economic setbacks and natural catastrophes, and the effects of these were exacerbated by the growing inability of the Awami League to govern effectively. In 1975, the Sheik was assassinated, and General Ziaur Rahman, Commander-in-Chief of the armed forces, took control of the government after a brief caretaker period.

Presidential and parliamentary elections were held (in 1978 and 1979 respectively). Zia's Bangladesh National Party won some 60 per cent of the seats in the 1979 election. Zia was assassinated in May 1981. Justice Abdus Sattar was elected as President later in that year, but in 1982 his government was dismissed by the armed forces, which continue at the time of writing to rule the country under the leadership of the Chief Martial Law Administrator, Lieutenant-General Ershad.

The Agriculture Sector

Agriculture dominates the Bangladesh economy, and one of the few certainties about the nation's future is that this will continue to be the case. About 75 per cent of the employed population work in the sector.

Agricultural output in this area depends on a rather delicate relationship between adequate and timely precipitation, and critical levels of inundation. This balance is easily disturbed, and production of rice padi and jute in particular are highly sensitive to disturbances. Crops are grown in in complex rotational patterns which vary greatly between regions, and between areas according to elevation, soil and other localised factors.

Rice is the predominant crop in the sector. Almost 80 per cent of all cropped acreage is allocated to it, and it is responsible for something in the order of 30 per cent of the entire GDP of the country. Bangladesh is the fourth largest rice producer in the world, after China, India and Indonesia. Rice accounts for some 85 per cent of the total calorific food intake of the country. In spite of its importance in the economy, rice yields in Bangladesh are low: in 1978-79, some 10 million hectares of cropped area yielded about 12.7 million metric tonnes of the grain - i.e. about 1.3 tonnes per hectare. This is less than yields in Java, and well below yields obtained in such places as Malaysia, Japan, Egypt and Italy. Overall production of rice has been rising but, as will be seen from Table A.1 below, increases in production have been well below the rate of increase in population and only averaged 1.7 per cent per annum during the 1970s.

Table A.1: RICE PRODUCTION IN BANGLADESH

Period	Per annum production (10^6 mt)
1960-61 - 1964-65	9.9
1970-71 - 1969-70	10.9
1970-71 - 1974-75	10.9
1975-76 - 1977-78	12.5
1979-80	13.5

Sources: Faaland and Parkinson (1976), Table 11.1; Statistical Yearbook of Bangladesh (BBS 1979); World Bank (1980).

There are a number of reasons why rice production has not increased as rapidly as had been hoped in Bangladesh. The irrigated winter rice crop, <u>boro</u>, has performed most poorly in recent years - this is partly due to competition for land by wheat (see below), but also, in 1978-79, for example, disruptions in delivery of diesel fuel and electricity for irrigation pumps occurred at a critical period in the growing season. Government output price support schemes have apparently not prevented discounting of high yielding variety grains on the markets, which affects the market for traditional varieties. Heavy subsidisation of agricultural inputs (especially for fertiliser and irrigation) may be an inefficient means of increasing output because of the heavy drain on Government resources, which otherwise might be used to more efficient ends in improving the input delivery system.

Jute is the second most important crop in the Bangladesh economy, and is the most important cash crop. It remains a major export earner for the country. Production in 1978-79 was 6.7 million bales - the highest since independence and well above the figures for 1977-78. In 1979-80, the crop fell to 6 million bales, partly because of the drought in 1979, but also because poor prices caused farmers to reduce acreages below the 900 000 hectares or so established in the previous year. The quality of the crop was also down, and this led to a further lowering of prices. The export market for jute is a somewhat precarious one: Burma and Thailand are competitors, and, particularly for lower quality grades seem able to best Bangladesh on price. The market is also subject to the vagaries of industrial unrest in the Indian jute mills - stoppages there not only reduce Indian demand for raw jute, but also increase its own deliveries of the unprocessed fibre on the international market. According to the World Bank (1980), closing stocks of jute in Bangladesh in 1976-77 (one of the better export years) were only 0.2 million bales, whereas in 1979-80, they were estimated at 2.94 million bales - this is more than the country could expect to export in a full year. The Government has attempted to ameliorate the farmers' situation somewhat, by raising credit availability to them, and by purchasing the crop. This has not cleared the stock problem, however, and it seems that farmers did not in fact receive the full Government price for the product.

One matter of concern in the move away from jute because of adverse market conditions, will be the effect on landless laborers: according to the Bank, jute requires about twice as much labour per unit area as rice cultivation.

A relative newcomer to the agriculture sector in Bangladesh is wheat. In 1970, only about 90 000 metric tons were produced, but by 1979-80, this had reached almost 1 million metric tons. Some problems of seed availability and crop disease are apparent, but overall the wheat crop is proving most beneficial. Wheat grain is higher in protein than rice. The crop can be grown over the dry season with lower irrigation requirements and it is more robust in terms of its reaction to water stress. To some extent, is a complementary crop: it can be grown in some areas in the off-season.

The tea industry is an important source of foreign exchange in the agriculture sector. In the early part of the country, there were some 250 gardens in the hilly north eastern area of the century. At present, about 150 remain, a total of 40 000 hectares of plantation. Yields for tea seem to be fairly low: about 700 kg/ha, compared to 1100 kg/ha in India, and 1000 kg/ha in Sri Lanka. Unseasonal downpours in the tea growing area are a problem, because they damage the crop significantly. Presently, the industry faces financial and other market problems, and its future is uncertain. The Government and the British aid agency are currently investing quite large amounts in refurbishment of the industry.

Other cash crops grown in Bangladesh include sugar cane (about 5.5 million mt a year), tobacco (40 000 mt) and oil seeds, with smaller crops of rape, mustard and pulses. Some re-expansion of the cotton crop, which used to be quite large before its decline in the 20th century, is presently being undertaken.

We can obtain an overall perspective on the agriculture sector for Bangladesh from the following table:

Table A.2: AGRICULTURAL STATISTICS FOR BANGLADESH

	1969-70	1976-76	1977-78	1978-79	1979-80
Population (million)	67.3	79.9	83.7	85.6	87.8
Total area of food grains (10^6 ha)	10.44	10.48	10.22	10.35	
Net(a) production of food grains (10 Mt)	10.443	11.682	11.985	11.914	12.116(c)
Total area of jute (10^6 ha)	2.46	1.28	1.81	2.05	
Production of jute (1000 bales)	7 171	4 248	5 359	6 443	
Exports of raw jute (1000 bales)		2 350	1 670	1 960	2 000(c)
Total area other crops 10^6 (ha)	1.86	1.66	1.67	1.68	
Foodgrain requirement(b) (10^6 Mt)	11 911	12 846	13 420	13 735	14 097

(a) Allowing 10 per cent for seed, feed and wastage. (b) Based on the Government target of 0.44 kgs/caput/day. (c) Projection.
Sources: World Bank (1980); Bangladesh Bureau of Statistics (BBS 1980).

According to the World Bank, the potential for gains in productivity in agriculture in Bangladesh is considerable: only 15 per cent of cultivable land is irrigated at present; the area under improved varieties of the major aus and aman rice crops is only 10 to 15 per cent of total cropped area; and fertiliser use by farmers is still well below optimal levels, even though it has more than doubled since 1972-73.

Obviously enough, from the above figures, productivity is going to have to rise more quickly, if the gap between production and consumption is not to widen further, necessitating even larger imports of food into the country. The WCARRD Review provides the following trend figures:

Table A.3: TRENDS IN AGRICULTURAL PRODUCTION AND IMPORTS FOR BANGLADESH

Item	1960-61 - 1969-70	1960-61 - 1976-77
	%	%
Cereals production	2.4	2.6
Fish production	1.8	1.0
Imports of cereals	9.8	7.3
Availability of cereals	1.8	1.6

Source: Table 2, WCARRD Review.

The final figure in this table suggests that per caput cereals availability has fallen by about 1.2 per cent a year since 1960-61: this trend would continue at about this rate if 1979-80 data are used.

The WCARRD Review also calculates some figures an annual rates of change of labour employment in various sectors of the rural economy, and derives a figure of 1.2 per cent a year since 1960-61. The agricultural labour force (or the potential labour force) would have grown at something in the order of 2 per cent a year during this period, based on working age cohort data in the demographic statistics. Not surprisingly, given this, real wages in the sector have fallen to very low levels:

Table A.4:   REAL DAILY WAGES FOR RURAL LABOUR: Taka/caput

1954-58 (annual average)	1.95
1964-68 (annual average)	2.15
1974	1.42
1975	1.28
1976-77(a)	1.27
1977-78(a)	1.28

(a) Deflators from inflation figures in official Budget Records (1972 = 100).

Sources: WCARRD Review Table 4; Economic Indicators for Bangladesh (BBS 1980).

These general figures speak for themselves. Rates of increase in production, income and employment in agriculture have lagged far behind population growth. Improvements in technology have been slow, not least because allocations of development capital to the sector have been disproportionately low. In all these respects, the sector typifies the general problem of urban bias and sectoral imbalance as discussed in Chapter 2. Major changes in Bangladesh society, and in financial allocations to sectoral development, are going to be needed to rectify the situation. The former, inevitably, will take time. The latter, on the other hand, can be altered to some extent from the top and could act as a catalyst of economic change.

The Industry Sector

According to World Bank (1980) figures, manufacturing industry currently accounts for 8 per cent of GDP in Bangladesh. However, more than 60 per cent of exports are manufactured goods, and some 350 000 people are employed in medium and large scale industry, whilst an estimated 4-5 million are employed in small scale industry.

Most major industries are under public ownership in Bangladesh, having been nationalised after the war in 1971. Jute goods; textiles; chemicals; and pulp and paper manufacture are all in this category. Some 70 per cent of all manufacturing industry is owned by the large government corporation, and about 85 per cent of all industry investment funds is utilised in them. Despite gradual improvement in the industry sector overall since Independence, many publicly owned industries have continued to incur heavy losses, and production problems are serious in most. Average capacity utilisation, although rising, had only reached 65 per cent in 1978-79.

Table A.5 below provides some basic data on the performance of the industry sector in Bangladesh:

Table A.5: PRODUCTION AND EMPLOYMENT FOR SELECTED INDUSTRIES IN BANGLADESH

Industry	Unit	Production			Employment	
		1969-70	1977-78	1978-79	1969-70	1977-78(a)
Jute	'000 mt	560	547	501	160 090	285 000
Cotton yarn	10^6 kg	48	49	44		
Cotton cloth	10^6 kg	55	77	78	58 800	167 000
Steel ingot	'000 mt	54	111	124	14 300	16 700
Paper	'000 mt	75	61	65	5 000	16 050
Electricity	10^6 kwh	1 913	-	-	-	-

(a) These figures derived by applying employment index figures (Table 5.24 SYB 1979 BBS) to 1969-70 figures. Note some disagreements with World Bank figures: the Bank gives a figure of 170 000 for employment in the jute industry in 1978-79, and a figure of 76 000 in the cotton textile industry for 1978-79.

Sources: 1979 Statistical Yearbook of Bangladesh (various tables); Second Five Year Plan, Table 13.2; World Bank (1980), Table 8.2

These figures do not include data on the small scale industry sector. According to the World Bank, the growth of this subsector is constrained by access to credit, and in 1978-79, it was expected to expand by only 2.5 per cent, compared to 6 per cent in the medium to large scale sector for that year.

Private sector activity in the industry sector remains low. The Government now favours encouragement of the private sector, and has sold off a number of public enterprises. Presently, however, many administrative discouragements to private entrepreneurs remain, and there does not appear to be widespread acceptance of the Government assurances on nationalisation policy as yet. There is, moreover, an absolute constraint on the level of private investment funds available, unless much more liberal lending policies are initiated by Government. Certainly there would not appear to us to be much chance of buyers for some of the large public corporation enterprises being found, even if those enterprises did exhibit some prospect of becoming profitable. Recently the Government has sought to encourage foreign investment, and is in the process of establishing a duty - free zone near the Chittagong Port for this purpose.

The Economy as a Whole

Table A.6 below gives an overview of the Bangladesh Economy through the decade of the 1970s.

Table A.6: GROSS DOMESTIC PRODUCT OF BANGLADESH: IN MILLIONS OF TAKA AT 1972-73 PRICES

	1972-73	1977-78	1979-80 (estimate)
Agriculture	28 830	34 540	36 680
Manufacturing	5 200	5 130	5 850
Construction	1 840	2 680	3 720
Power and gas	150	410	570
Housing	2 360	2 810	3 030
Trade, transport, services	11 650	15 320	17 530
Total	50 030	60 890	67 200

Source: Table 1.2, Second Five Year Plan (Planning Commission 1980).

These figures yield an aggregate GDP growth rate of 4 per cent a year between 1972-73 and 1977-78. On the basis of a 2.8 per cent a year rate of change of population, this implies a 1.2 per cent a year growth in per caput income. Even this fairly disappointing figure may be an over-statement of the underlying growth rate, in that the base year, 1972-73, was an exceptionally poor year for Bangladesh in most respects.

There is no doubt that the poorest group in society is bearing a disproportionately large share of the poor performance of the economy. Per caput expenditure in real terms of the bottom 40 per cent of the rural population in 1976-77 was only two-thirds of that in 1963-64. The World Bank (1980) defines a 'hardcore' poverty line as being an average food intake of 1805 calories per day (this is 85 per cent of the minimum amount recommended as necessary by nutrition exports), and calculates that some 60 per cent of the rural population, and 40 per cent of the urban population fall below this line at present. The Bank acknowledges that there is no proven firm relationship between calorific food intake and human welfare, but points out that intakes have declined over time in Bangladesh - seriously so for lower income groups. The Bank is in no doubt (and neither are we) that this trend, and that of growing landlessness, indicate intensifying poverty.

It is rather more difficult to obtain useful figures on the associated problem of unemployment. Faaland and Parkinson suggest that in 1975, 8 million of the total labour force of 28 million (i.e. 29 per cent) was unemployed, but they warn that this is only a rough estimate. Underemployment is high, but variable. It would seem to us that, allowing for underemployment, it is most likely that the effective unemployment of the labour force would certainly now exceed 30 per cent. Faaland and Parkinson provide some indicative calculations to show that on their population forecasts, and utilising fairly optimistic assumptions about the capacity of the major sectors of the economy to observe absorb new labour, this figure rises to 36 per cent by 2000 AD. We would not be surprised to learn that this figure has already been reached.

Inflation is also difficult to measure, in a country where much consumption and trade is based on barter, and where a welter of official price controls, ration systems and so on exist. Rahim and Uddin (1978) have criticised official price index figures in Bangladesh, which are based on purchases of the urban middle class. They claim this index understates trends in real costs of living index for middle income families in Dacca from June 1978 to June 1979.

As is well known, Bangladesh has an adverse balance of trade, and the trend has intensified:

Table A.7: IMPORTS AND EXPORTS FOR BANDLADESH: IN MILLIONS OF TAKA

	1955-56	1970-71	1973-74	1977-78	1978-79
Exports	104.1	125.1	298.3	718	900
Imports	36.1	147.5	732.0	1 822	1 499
Balance	+68.0	-32.4	-433.7	-1 104	-1 593

Sources: 1979 Statistical Yearbook for Bangladesh (BBS 1980); Economic Indicators of Bangladesh (BBS 1980); Far Eastern Economic Review, 22 June 1979.

Large inflows of development assistance funds have averted, so far, the quite serious foreign exchange and economic management problems that these figures would otherwise imply. Development expenditure in 1975-76 was about Tk. 9500 million. In 1978 the figure rose to Tk. 12 020 million, of which 79 per cent was foreign financed. The figure planned for 1979-80 was given in the Second Five Year Plan as Tk. 23 300 million: some Tk. 15 500 million was disbursed in this year by foreign donors. In 1980-81, the foreign assistance figure was expected to be Tk. 21 000 million, but there were at the time of writing some indications that donor commitments were being reduced from previously stated levels.

Appendix B

FOREST INDUSTRIES OF BANGLADESH

Paper Manufacturing and Pulp Products

Paper making in Bangladesh is completely dominated by the Bangladesh Chemical Industries Corporation (BCIC), a large Government run conglomerate which also controls some 30 industries in the fertiliser, pharmaceuticals, glass and ceramics, rubber goods, packaging and match-making sectors.

There are three paper manufacturing factories: one (Khulna Newsprint Mills (KNM) makes newsprint and mechanical print, the other two (Karnaphuli Paper Mills (KPM) and North Bengal Paper Mills (NBPM) manufacture printing and writing papers. The Sylhet Pulp and Paper Mill (SPPM) manufactures pulp, for use in other operations. Originally, all paper-making in Bangladesh except newsprint was to be based on non-wood local pulp furnish (with some imported long fibred woodpulp to obtain strength in the products). KPM was to operate on 100 per cent bamboo local furnish; NBPM was (and remains) based wholly on bagasse; SPPM was to use bamboo, jute-sticks and reed. In the case of KPM, bamboo supplies to the mill were not able to be maintained, and it now operates on a 40 per cent wood pulp furnish. The reed supply for SPPM has virtually disappeared, and bamboo for it is also in short supply.

Attached to the Karnaphuli Paper Mill is Karnaphuli Rayon and Chemicals (KRC) which products rayon and cellophane (dilphane) from a 100 per cent bamboo furnish.

Some basic economic data on the pulp and paper industries of Bangladesh are given in Table B.1.

Table B.1: ECONOMIC AND OUTPUT DATA: BANGLADESH PULP AND PAPER INDUSTRIES

	Unit		1976-76	1976-77	1977-78	1978-79	1979-80
Production	'000 mt	KNM	20.8	16.7	31.6	37.2	41.3
		KPM	14.2	19.4	23.1	22.6	23.2
		NBPM	3.1	4.1	6.5	na	na
		SPPM	-	-	9.3	12.1	12.0
		KRC	1.7	2.0	2.0	2.2	2.6
Domestic sales	'000 mt	KNM	11.8	14.9	17.5	21.3	20.0
		KPM	13.8	15.8	17.8	17.0	18.2
		NBPM	3.3	4.0	6.5	na	na
		SPPM	-	-	6.8	12.1	12.0
		KRC	1.3	1.6	2.2	2.1	2.1
Average price of domestic sales	Tk/mt	KNM	3 859	5 664	5 945	6 301	7 081
		KPM	7 429	9 691	9 932	10 695	12 574
		NBPM	na	9 538	9 813	na	na
		SPPM	-	-	5 125	5 699	6 921
		KRC(a)	24 339	36 360	43 062	53 055	59 590
Export sales	'000 mt	KNM	4.9	7.6	21.3	20.3	21.4
		KPM	0.5	2.3	3.5	1.7	4.1
		NBPM	na	-	-	na	na
		SPPM	-	-	5.2	-	-
		KRC	0.3	0.5	0.5	0.7	0.7
Average price of export sales	Tk/mt	KNM	3 612	3 962	4 078	4 372	5 878
		KPM	7 005	6 557	7 328	9 703	10 817
		NBPM	-	-	-	-	-
		SPPM	-	-	2 588	-	-
		KRC(a)	20 458	32 812	30 129	na	49 693

Continued on next page

Table B.1 (continued)

	Unit		1976-76	1976-77	1977-78	1978-79	1979-80
Production cost	Tk/mt	KNM	5 753	6 898	5 543	5 590	7 572
		KPM	11 455	10 033	10 086	10 933	13 374
		NBPM	na	16 268	14 008	na	na
		SPPM	-	-	-	10 612	11 264
		KRC(a)	55 136	48 820	60 725	63 937	68 665
Operating surplus	Tk10⁶	KNM	-56.6	-1.0	+14.7	+14.7	-45.4
		KPM	-56.2	-26.7	-30.5	-50.5	-33.6
		NBPM	na	-29.5	-28.1	na	na
		SPPM	-	-	-63.9	-58.8	-51.8
		KRC	-44.4	-30.8	-31.0	-16.4	+9.0
Output/ employee	'000 Tk	KNM	30.4	55.3	89.6	102.2	115.4
		KPM	29.8	48.9	57.4	54.9	77.8
		NBPM	na	na	85.1	na	na
		SPPM	-	-	42.6	58.6	70.6
		KRC	19.2	26.2	40.5	48.6	63.3
Labour productivity (total revenue/ labour cost)		KNM	6.3	10.0	13.7	14.2	12.3
		KPM	4.5	6.1	5.2	4.9	6.4
		NBPM	-	-	6.4	6.6	4.2
		KRC	2.8	3.5	3.6	3.6	4.1

(a) KRC sales prices and production costs based on weighted average of rayon and cellophane.

Sources: See Douglas et al. (1981).

Although not shown in this table, in every case these mills have recorded particularly rapid rises in imported input costs over the data period. In the case of KNM, the mill management estimates that foreign-derived imports account for 65 per cent of total costs. In the case of KPM, an estimated 50 per cent of production cost is spent on imports.

The economic outlook for these industries, on the basis of the data, does not seem promising. Those which sell appreciable volumes of output to export do so at prices well below cost. Those which rely on the domestic market sell at very high prices (certainly well in excess of imported equivalent prices), and yet continue to record operating losses. Capacity utilisation (not shown above) varies: in KNM, it has been relatively high, similarly KPM. In NBPM, however, it has never risen much beyond 50 per cent, and in SPPM never beyond 70 per cent.

Most of the established plant is old (KPM was commissioned in 1953; KNM in 1960), and units are far too small to attain current economies of scale. Yet, having been designed originally with the highly protected West Pakistan market (which evaporated after Independence) in mind, the installed capacity is well in excess of the needs of the Bangladesh market (even at the controlled, highly subsidised prices for which most output is sold).

The Khulna Newsprint Mill does have access to an export market (principally India, by virtue of transport cost advantages) albeit at prices well below costs of production. Although KNM do generate positive foreign exchange balances, the costs of so doing (if this is interpreted as the principal benefit of industries such as this) are extremely high for Bangladesh.

Nor are the employment benefits generated by the industries particularly impressive: even compared to the capital intensive norm for Bangladesh Chemical Industries Corporation operations, the pulp and paper group employs less than the average per unit of capital invested. In the case of KNM, it can be calculated (see Douglas et al. 1981) that the average amount actually paid to employees over the five year data period

1975-76 to 1979-80 was Tk. 34 000. But the cost of subsidising KNM over the same period, in per caput employee terms, was Tk. 121 000.

Manufactured Board

Bangladesh is involved in the manufacture of hardboard, particleboard and plywood. Hardboard is produced by the Khulna Harboard Mill (KHM), a BCIC operation which utilises Gewa raw material supplied by the logging program of the Khulna Newsprint Mill. Particleboard is presently manufactured at the Star Particle Board Mill (SPBM), another BCIC operation. The product in fact is based entirely on jute sticks - but it will be considered here because of what this type of product tells us about the market in Bangladesh: there is, at the time of writing, a wood based particleboard plant also due to go into operation.

There are seven plywood plants operating in Bangladesh. One, the Sangu Valley timber Industries (SVTI) is owned by the Bangladesh Forest Industries Development Corporation (BFIDC), a Government conglomerate of 20 enterprise, ranging from logging and extraction projects in the Chittagong Hill Tracts, through to finishing units for furniture and other products. The remaining units are privately owned. For present purposes, no important detail is lost by grouping the plywood units together, and this has been done in the data below.

Both the particleboard and plywood industries turn out a range of products. There is little point in showing averaged prices or production costs for these products (prices and costs for individual products are shown in Douglas et al. 1981). Here, the basic indications of economic viability can be deduced from operating surplus and other figures below, for these industries. For the ply industry, it is not possible to calculate overall domestic sales from the available data. In any event, for this industry it is obvious that sales follow production very closely.

Table B.2: ECONOMICS AND OUTPUT DATA: BANGLADESH MANUFACTURED BOARD INDUSTRIES

Item	Unit		1976-76	1976-77	1977-78	1978-79	1979-80
Production	'000 m2	KHM	1 264	1 589	1 858	1 765	1 598
	mt	SPBM	1 684	2 080	3 098	2 101	2 476
	'000 m2	Ply	697	780	1 500	1 000	1 200
Domestic sales	'000 m2	KHM	1 189	1 477	1 756	1 505	1 449
	mt	SPBM	820	1 543	1 561	1 692	1 750
Average domestic price	Tk/'000 m2	KHM	6 243	6 211	6 286	7 718	11 744
Export sales	'000 m2	KHM	149	177	130	279	158
	mt	SPBM	893	1 586	471	959	na
	'000 m2	Ply	–	–	–	–	–
Average export price	Tk/'000 m2	KHM	6 803	5 382	5 630	6 211	8 913
Production cost	Tk/'000 m2	KHM	7 287	6 620	6 372	7 642	11 173
Operating surplus	Tk 10^6	KHM	-0.7	-0.3	-0.1	-0.2	+0.5
	Tk 10^6	SPBM	-5.0	+1.5	-6.8	+0.5	na
	Tk 10^6	Ply(a)	-1.3	+0.9	+0.1	+1.17	+1.52
Output/ employee	Tk '000	KHM	33.3	40.8	46.4	52.8	73.0
	Tk '000	SPBM	12.2	29.1	22.2	32.7	na
	Tk '000	Ply(a)	20.6	34.5	48.2	56.8	59.6
Labour productivity (total revenue/ labour cost)		KHM	6.1	7.3	5.2	4.3	5.5
		SPBM	2.3	4.9	2.8	3.5	na
		Ply(a)	4.4	6.0	9.8	7.2	5.7

(a) Based on return from the three largest plymills, of the seven in production in Bangladesh.

Source: See Douglas et al (1981).

Although the hardboard plant of KHM is quite old, and is of poor design, KHM seems able to cover its costs of operation. The principal reason for this would seem to be KHM's ability to sell the great majority of its product to the domestic market, for prices which, by international standards, seem reasonable. Provided some control can be maintained over imported input costs, which for this industry have risen very rapidly over the data period, the industry seems reasonably viable, financially. It is interesting to note that a second hardboard plant owned by BFIDC was closed down at Chittagong several years ago. The reasons given for this were quality and costs problems, but it may be that these were counterveiling a basic shortage of effective demand for the output from both operations.

The particleboard mill (SBPM) is in a much more parlous condition, financially. The overall capacity of this mill is 4000 tons — well above current production. The output produced is of fairly low quality, and is extremely highly priced (around Tk 54/m^2 in 1979-80, compared to international parity prices of say Tk 30-40/m^2 for this product in that year). This is the more disturbing because another particleboard mill, operated by BFIDC, is shortly to commence operation, based on the wood raw material resource of the Chittagong Hill Tracts (see our general remarks on this subject of the beginning of this section). If that mill is able to obtain sufficient raw materials, and operate efficiently, its output will probably displace that of SPBM from the domestic market — in which case the only option would be to close it down, or export its output for even greater operating losses than currently apply.

Most plywood made in Bangladesh goes into tea-chests, and the future of the plywood sector therefore depends largely on the fortunes of the export tea industry (see Appendix A above). The sector faces some quite severe raw material problems: the Civit, Semul and Kadam logs it requires for its production are often difficult to obtain. Nevertheless at present the existing operations seem able to maintain financial viaibility. The Ruby Mill, which ceased production in 1979 (explaining the drop in output figures in Table B.2 above under ply for that year) has recommenced operation in 1981.

Urban Sawmilling

Fewer data are available on the sawmilling sector of Bangladesh: the great majority of it is carried out in very small, often ephemeral rural units about which virtually nothing is known (see next Section). Even in the urban areas, where sawing is done in small permanent mechanical mills, very few data are available. In Douglas et al. (1981), a procedure is described where the majority of sawmills in the Dacca metropolitan area were contacted with a brief questionnaire. This, plus the official data on the larger Government (BFIDC) mill at Kaptai, in the Chittagong area, constitutes the totality of information presently available.

Some 140 sawmills exist in the Dacca area and from questioning and follow up research it is reasonable to suppose that virtually all of their output is used within the metropolis, and that this constitutes the great majority of <u>total</u> sawnwood usage in the metropolis. The mills produce an average 540 cubic metres of sawn output per annum, which amounts to some 75 000 cubic metres supply annually to Dacca. Most of the mills reported irregularity of log supplies, and electrical power failures, as their major problems. All were able to sell all their sawn output, and appeared financially viable.

The large BFIDC sawmill at Kaptai has an annual capacity of around 14 000 m^3 sawn output. Its basic design was as a primary breaking-down unit: output was then to be transferred to smaller operators for finishing. This <u>modus operandi</u> has not eventuated: the mill is forced to cut a multiplicity of small, irregular orders (many of these emanate from different Government departments, which do not standardise sizes for large volume items). Presently, the mill output is less than 4500 m^3 a year, and it has generated substantial operating losses over recent years. However, given some management and marketing re-organisation, and possibly re-location of the large seasoning kilns attached to the mills there would seem to be no reason why the full capacity output of this mill could not be profitably sold within Bangladesh. Demand for sawn timber is extremely high, and very high prices are paid for it.

Match Factories, Textile Accessories, Pencils etc.

As estimated in Douglas et al. (1981), current output of matches in Bangladesh is about 45.7 million gross boxes in which uses some 57 000 m^3 roundwood to manufacture. The match making industry is carried on in a number of small units around the country, some of which the Government (BCIC) owned. Although not a heavy user of wood, the industry is a large employer - in 1979-80, for example, it employed almost 6000 workers - more than the two large paper mills (KPM and KNM) combined. Most of the factories have old plant, and repairs and maintenance are extremely costly. However, most of the units examined were profitable, and positive operating surpluses for the sector as a whole have been recorded in recent years.

There are two pencil factories in Bangladesh, and 12 registered manufacturers of bobbins and other textile mill accessories (although it is known that many more operators than this are currently involved in supply). These industries are not significant consumers of raw material, and little is known of their operations.

Rural Sawmilling

As noted earlier, no detailed information is available on this sector. Certainly, we can, from observation, safely exclude it from the 'large-scale' definition of the previous section. There are some mechanical mills operating in the rural area - there are a great many hand - operated pitsaws.

From some general studies of transport into major cities of logs and wood products, it seems that little sawn material moves from rural mills into urban areas. To the extent that this is so, it allows us at least to estimate the overall size of the rural sawmilling sector: the rural wood consumption survey described in Douglas (1981) estimated use of sawn material in rural construction, furniture, implements and so on. We can safely assume that this material was supplied by the rural sawmilling sector, and that it should represent more or less the full output of the sector. On this basis, output of the sector is 830 000 cubic metres a year, estimated in 1979.

Appendix C

CONVERSION OF PLAIN LAND FORESTS OF BANGLADESH TO FAST GROWING PLANTATION

It is apparent that the plain land forests of Bangladesh are currently underutilised. And, as argued in subsection 4.3.5 of this book, the area where this resource is concentrated is close to the north-western region, where a critical wood shortage – particularly for domestic fuel use – is in the offing.

Some of the deficit could be met by interregional imports (although other regions are by no means in surplus). Some of it could be met by substitution and technical change, but these are unlikely to entirely overcome it. If it is not to come from continued depletion of current homestead forest growing stock (which <u>cannot</u> continue at present levels for long), or from further reductions in already minimal per caput fuel usage levels, then a resource will need to be created.

A suitably selected and developed fast growing fuelwood/structural wood species, growing in the tropical environment of the Bangladesh deltaic plain, should be capable of returning a volume growth rate of 25 m^3/ha a year(1) at maturity. Maturity (for fuel cropping purposes, at least) should be attained in six or seven years. Eucalyptus <u>camaldulensis</u> (northern province), which is generally thought to be a suitable species for Bangladesh, has certainly attained this sort of growth level in various parts of the world (including the Sub-Continent), and there is a significant number of other species of tree which have attained far higher rates of growth than this. In addition to this log volume, trees will also supply litter fall (small branches, leaves etc.) throughout their life time and, obviously, this material has fuel value in Bangladesh.

On these figures, a (conservative) projected annual smallwood deficit of 2.5 million m³ in the northern and central area of the country (see Douglas 1980) could be supplied permanently from around 100 000 hectares of mature (for those purposes) plantation. This is about the area of land the Government currently has available under <u>Sal</u> associated forest in this area. Not all of the land may in fact be currently forested – some of it is probably in a degraded condition, and some is probably encroached by farmers. But at least the above orders of magnitude show that significant inroads into the smallwood deficit could be made by a rapid program of plantation of currently utilisable land in this resource.

It has been argued in subsection 4.3.5 that the use of plain land forest areas <u>could</u> be used as an effective means of distributing income on a permanent basis towards improverished and landless groups – and this is a necessary objective in land use programs in Bangladesh. Any analysis of a conversion option should, therefore, have implicit in it some provision for this. Assuming that a landless family or family sized group of, say, 5 persons, could obtain an adequate living from 1 hectare of fast growing plantation (allowing for some undercropping with agricultural or fodder crops), conversion of the available <u>Sal</u> areas could therefore directly involve half a million people in forestry, and the considerable income and output multiplier effects that could be expected would probably double this impact. At a maximum, then, successful operation of such a program could involve a very large number of people in income bearing activity on land which currently is earning very little, in national terms.

An Indicative Analysis of Conversion

Such arguments as these alone justify <u>some</u> form of conversion option. However, it is necessary to perform more quantitative analysis to estimate how such a program would perform, in terms of returns to capital and land. Whilst actual representative data for such a project do not exist, it is possible to use some reasonable indicative figures. Assume:

(1) Payment of Tk 1500/- annually to a selected family (or group) to establish and maintain tree cover over 1 hectare of land, until the year of harvestable maturity.

(2) Fast growing trees will reach a mature output of 25 m^3/ha a year in year 8 after establishment.

(3) Such plantations will produce an additional 6 m3/ha a year leaves, twigs, bark etc. from year 4.

(4) The initial clear felling of <u>Sal</u> from selected sites yields 35 m^3/ha.

(5) The Forests Department charges itself a cost of Tk 20 000/ha for land to be converted.

(6) The cost of land preparation, seedlings, establishment and so on (<u>net</u> of uncosted tenant family labour) is Tk 7400/ha (Tk 3000/acre).

(7) The price of output is based on a fuelwood price at market of Tk 540/m^3 (Tk 25/md), from which is subtracted Tk 100/m^3 for handling and delivery costs (<u>net</u> of uncosted tenant family labour).

(8) At year 8 (maturity), the tenant family is no longer paid the Tk 1500/- a year retainer, but instead receives 20 per cent of all volume output from the 1 hectare of plantation, in addition to whatever undercropping or grazing produce is available through their own efforts. These assumptions allow calculation of an internal rate of return to the investment; on the following basis:

  (i) To obtain a Tk 1500 a year retainer payment, invest Tk 10 000/- at 15 per cent a year in year 0.

  (ii) Value of initial Sal fellings 35 m^3/ha (at Tk 440/m^3 = Tk 15 400.

  (iii) Establishment cost etc. Year 1 = Tk 7400/ha.

(iv) Initial land cost Year 0 = Tk 20 000/ha.

(v) Returns year 0 - 3 = 0.

(vi) Returns years 4 - 7 = 6 m^3/ha of litter fall valued at, say Tk 10/md (Tk 214/m^3) = Tk1287/ha/year.

(vii) Returns year 9: return of Tk 10 000/- retainer investment; plus value of output of fuelwood 25 m^3 at Tk 440/m^3; plus value of 6 m^3/ha of litter fall at Tk 214/m^3; less tenant's 20 per cent of annual fuelwood crop Tk (0.2 x 25 x 440) = Tk 20 087.

(viii) Returns year 9 onwards: value of output of fuelwood, less tenant's Tk (0.8 x 25 x 440); plus value of 6 m^3/ha of litter fall at Tk 214/m^3 = Tk 10 087 a year.

On this basis, applying the usual IRR formulation, this investment would generate an internal rate of return to the investor - the Forests Department - of 18.2 per cent, in addition to generating income from fuelwood of Tk 2200 per tenant family from year 8 onwards, in addition to whatever they would earn by undercropping.

2. Comparing the Agriculture Alternative

It is valid, especially in a food deficit country such as Bangladesh, to inquire what returns might be generated from the use of such land for agricultural purposes - there being no immediately obvious means, in the very straitened circumstances of Bangladesh, for making meaningful trade off decisions on the basis of agriculture and forest output utilities. The land upon which Sal forests currently grow is slightly elevated, and tends to be somewhat more leached than surrounding agricultural land. It is rational, when using productivity factors based on other land, to account for this.

In Douglas (1980) an agricultural land use option is evaluated. That option is an optimistic one to compare, in that it is based on a fairly highly productive double-cropping regime of jute and transplanted Aman rice padi. It is highly unlikely that this intensity of production could be obtained on Sal forest sites, and the net return of Tk 5000/- per hectare a year from this agricultural regime is therefore reduced in this analysis to Tk 4000 - per hectare per year - which is probably still somewhat optimistic.(2)

Using the same costs assumptions as for the plantation alternative (except that no initial investment of Tk 10 000/ha is required to generate incentive payments from year 1), this option yields an IRR of 18.9 per cent.

3. Summary and Conclusions

Conversion of Sal forest to fast-growing plantations would, on the above calculation, give an IRR of 18.2 per cent - comparable, within the scale of accuracy possible here, to the rate of 18.9 per cent generated under agriculture. For a number of reasons, the plantation option may be preferable, given the closeness of outcome of the IRR figures:

(i) The plantation returns figures are fairly conservative: it seems apparent from observation that a fuelwood price of Tk 25/md is probably already an underestimate of the market price for this product. Moreover, no margin for sawlog size and quality material has been allowed in this analysis.

(ii) No benefit is calculated under the plantation option for undercropping (which could yield considerable income). This form of production is not possible under the pure agriculture alternative.

(iii) It is probable that forestry is a better form of land use for the areas in question, in terms of environmental factors of: water regime; erosion control; utilisation of natrients, in top-leached, elevated sites; soil restoration and enrichment. Also it may be preferable in view of the fact that many of the current encroachers in forest land are, of history and tradition, forest dwellers.